TEACHING PRESCHOOL
MATH

TEACHING PRESCHOOL MATH

Foundations and Activities

Anthony C. Maffei, Ph.D.
Patricia Buckley, M.A.

The William Paterson College of New Jersey, Wayne

HUMAN SCIENCES PRESS

72 Fifth Avenue   3 Henrietta Street
NEW YORK, NY 10011 ● LONDON, WC2E 8LU

Printed in the United States of America
123456789   98765432

Library of Congress Cataloging in Publication Data

Maffei, Anthony C
    Teaching preschool math.

    Bibliography: p. 167
    Includes index.
    1. Mathematics—Study and teaching (Preschool)
I. Buckley, Patricia, joint author. II. Title.
QA135.5.M287   372.7   LC 79-27448
ISBN 0-87705-492-4

CONTENTS

ACKNOWLEDGMENTS

This book could not have reached this state had it not been for several people.

Susan Mitchell, Director of Kiddie Korner nursery school, graciously offered us a place, as well as ideas, for testing some of our activities. She, Barbara Marieni, and the teaching staff were extremely instrumental in organizing those courteous and adorable children for picture taking.

The drawings and sketches are from the talented hands of Florence Brown.

Our contact people at HSP have been most helpful and patient.

And finally, the secretarial skills of Doris Barlow converted our frequent scribbles into professional manuscript form.

A. M.

P. B.

Part I

FOUNDATIONS

Chapter 1

INTRODUCTION

GROWTH OF PRESCHOOL PROGRAMS

Statistics show that preschool programs are currently growing in number at a faster rate than they did 10 years ago. There are probably several reasons for this. One is probably the women's liberation movement in the United States. Proponents of the Equal Rights Amendment have helped women free themselves of their stereotyped images as housewife and homemaker. Another cause could be the change in our economy—with the steady rise in the cost of living, one income, in many instances, is not sufficient to pay all the bills.

The momentum, therefore, at the present time is toward sending 3-, 4-, and 5-year-olds to preschool. This, as Berson and Sherman point out in a recent article (1976), leaves today's educators faced with the ironic dilemma of a declining birth rate and a rising preschool attendance:

> Two antithetical conditions currently exist in American education. There is a critical shortage of teachers for three-, four-, and five-year olds on the one hand and a critical shortage of elementary and secondary teaching jobs on the other.(p. 143)

The authors' figures show that 45.2 percent of the children in the 3- to 5-year-old age group are presently in preschool programs, with about 8 million students currently in need of such programs. The American Federation of Teachers (AFT) has been urging the federal government to spend 20–40 billion dollars to supply free and voluntary care for all preschool children, with the purpose of combining the "job needs of its members with the day care needs of the nation."(Levinson, 1977, p. 67.) Some industries have been creating preschool programs to serve the needs of their employees, while institutions such as hospitals and universities have been providing facilities for the children of both their professional and service staffs (Hess & Croft, 1972).

Children from economically, socially, and educationally disadvantaged environments are in real need of preschool programs, as "about three-fourths of all diagnosed cases of mental retardation result from adverse environmental conditions in early childhood, particularly lack of preschool development experiences," and "recent research indicates that early childhood experiences are the most important factors causing this milder form of retardation" (Phi Delta Kappan, 1978, p. 369).

QUALITY OF PRESCHOOL PROGRAMS

Preschool education has various names, including, Head Start, day care, and nursery school. The educational quality of all of these varies from state to state, and even from locality to locality. Some states require that their teaching personnel be certified, while others have no such requirements. As a result, the educational standards and quality of preschool programs vary tremendously and it is not uncommon to find children in one preschool setting watching television all day, while in another, not far away, the children are engaged in guided play activities.

This passage, from a text written over ten years ago, by James Hymes (1968) still reflects the educational quality of some preschool programs:

> People are fond of saying that nursery-kindergarten education is the best, primary education is the next best, and so on up the list until you come to graduate education which allegedly is the worst of all. If there must be graduations, this is as it should be. Early education ought to be very, very good. Often it is. Sometimes it is not. Many programs are too pushy. They are too narrow; they hold the wrong goals. Other classrooms are too passive. They challenge their children too little. They let their youngsters drift and wander. (p. 4)

Probably one of the most important reasons that preschools vary so much in what they offer their students is that though educators have defined what constitutes the elementary school curriculum, they have never decided what makes up a preschool curriculum. As a result, for a long time, there was a tendency for preschool educators to put together (or not put together!) a preschool program without paying much attention to the cognitive development of the children within this age level. As David Elkind points out, part of the blame rested with the government:

> Government-sponsored preschool programs made it possible for professionals to try out formal educational programs with young children. At least some of these programs . . . were undertaken without sufficient consideration for the human rights of the children involved. (p. 55)

If it were not for Jean Piaget's detailed descriptions of the intellectual characteristics of the preschooler, educators might still be groping for an appropriate preschool curriculum.

PARENTS AND THE PRESCHOOLER

Different parents have different feelings and reactions when faced with the possibility of sending their 3- , 4- , and 5-year-olds off to school. Some parents feel that they should keep their child at home, while others react quite differently:

> There are the status pressures to have children who read early and do well in school. There are the personal pressures to give their children the educational advantages they never had. Finally, and most important, there are the guilt pressures suffered by mothers who feel that they should be taking care of their preschoolers instead of putting them in a preschool or day care center. (Elkind, 1971, p. 55)

The last type of pressure is definitely the most troublesome and Elkind believes that parents, especially mothers, should "come to appreciate that the 'maternal instinct' is a fable and that young children suffer no emotional trauma by spending their days in a nursery or day care center" (p. 55). Actually at this age level children need the experience of socializing with their peers in an environment where doing and making are the activities of chief concern.

ABOUT THIS BOOK

The major portion of the first part of this book is devoted to studying the cognitive characteristics of the preschooler as described by Piaget. The second part presents a preschool math curriculum based mostly upon these theories and activities that should provide the preschooler with a rich background for first grade work.

In their book, *Mathematics Begins,* the Nuffield Mathematics Project described the reasons that this book was written:

The background of experience of the child determines the starting point when he comes to school. If he has not been fortunate enough to have enjoyed a rich and varied set of activities in these early years—if he has not been able to discuss these with someone who uses language with flexibility and imagination, then these opportunities must be made available in school as a first priority, for on such a foundation does his future development depend. (p. 5)

We hope that this book will provide preschool educators with ideas that will contribute to that foundation.

Chapter 2

BEGINNING OF PREOPERATIONAL THINKING

Background Ideas

Jean Piaget is recognized today in many academic areas as the leading expert in the thinking characteristics of young children and adolescents. Piaget divided the development of intellectual thought into four general stages:

1. the sensorimotor period, 0 - 2 years old;
2. the preoperational period, 2 - 7 years old;
3. the period of concrete operations, 7 - 11 years old; and
4. the period of formal operations, 11 - 15 years old.

According to Piaget, these stages of development must be sequential for normal intellectual development to occur. That is, a child must encounter enough learning experiences within each stage in order to proceed to the next. Since the ages for each developmental stage are approximate, it is quite possible, for example, to have some 8- and 9-years-olds reasoning on the preoperational level.

Chapter 3 will examine in detail the preoperational thought characteristics of the preschool child, however, in this chapter, we will discuss briefly some basic characteristics of the preceding period—the sensorimotor. The factors which affect the four stages of intellectual development will also be introduced, as will the emergent characteristics of preoperational behavior.

The sensorimotor stage is characterized as nonverbal. In the beginning, Flavell (1963) states that the infant shows "little other than a few uncoordinated, reflexlike activities, such as sucking, tongue movements, swallowing, crying, gross bodily activity and the like" (p. 89). Toward the end of this stage the child shows signs of thinking internally and finds "new means not only by external or physical groping but also by internalized combinations that culminate in sudden comprehension or insight" (Piaget & Inhelder, 1969, p. 11). As an example, Piaget cites the dilemma faced by a child attempting to get a thimble from a slightly opened matchbox. After several futile attempts, the child stops to examine the situation, then slips a finger into the crack of the box, opens it, and picks up the thimble. The same child at an earlier period of development might have thrown the matchbox around the floor in order to get at the thimble.

The child's search for an appropriate means to an end is characterized by Piaget as the beginning of internal thinking and verbal behavior (Copeland, 1979), and verbal behavior marks the beginning of the preoperational stage.

There are, according to Piaget, four factors that have a direct bearing upon whether or not a child does progress from one stage of intellectual development to the next. They are: organic growth (maturation), experience, social interaction or transmission, and equilibration.

Organic growth deals with the physical growth of the individual "especially maturation of the nervous and endocrine systems" (Copeland, 1979, p. 29).

Experience, the second factor affecting intellectual development, generally deals with a child's exposure to the environment. Giving preschool children experience, as we shall see later, means getting them involved in working with objects such as blocks, rods, buttons, and sticks. Failure to allow children to explore mathematics through these manipulatives can delay development of their ability to conserve length and number.

The third factor, social interaction, which implies that all humans learn by verbal and nonverbal interaction with their peers, parents, and teachers, makes it obvious that most preschoolers need help and direction from the people around them in order to learn successfully.

As Copeland points out (1979), equilibration, the last factor, is very basic to Piaget's theory of intellectual development and is a unique concept to this theory:

> The process of equilibration is an active process involving a change in one direction being compensated for by a change in the opposite direction. Children of less than seven are usually unable to reconcile apparent contradictions such as that when one stick is placed above another it looks longer, yet placed side by side the sticks are found to be of the same length. Older children are able to accommodate to this information—that the position a stick occupies does not determine its length. A new mental "equilibrium" has been established. (p. 30)

For Piaget, the term equilibration "is used in the sense that it is used in cybernetics—processes with feed back and feed forward, of processes that regulate themselves by progressive compensation of systems, in a sense like an electronic computer" (Copeland, 1979, p. 33).

LANGUAGE AND THINKING

Perhaps the single most important characteristic marking the preoperational stage is the emergence of language. The use of words allows children a great deal of mobility in communication with others that, up to that point, they had been unable to experience. In the sensorimotor stage, children's intellectual or thinking ability is limited to their actions upon objects. If they want to make an idea known to a parent or playmate, they have to do it by pointing to objects or by taking the person by the hand to show what they mean.

Therefore, although the child is already thinking in the sensorimotor stage, prior to language acquisition, it is a different form of thinking than the logical thinking employed in the later preoperational and concrete operational periods:

> Intelligence actually appears well before language, that is to say, well before internal thought, which presupposes the use of verbal signs (internalized language). It is an entirely practical intelligence based on the manipulation of objects; in place of words and concepts it uses percepts and movements organized to "action" schemata. (Piaget, 1968, p. 11)

With the development of language, the children's thought processes are freed from dependence upon objects and movement and they are able to go through daily life experiences at a faster rate than before. According to Wadsworth (1971), this new freedom directly affects their intellectual development:

> When language develops there is a parallel development of conceptual abilities that language helps to facilitate, probably because language and representation permit conceptual activity to proceed more rapidly than sensorimotor operations do. Language development is seen as a facilitator of cognitive development. (p. 69)

The emergence of verbal behavior improves the range and rapidity of thinking in several ways when compared to sensorimotor thinking. Since "sensorimotor patterns are obliged to follow events without being able to exceed the speed of the action, verbal patterns, by means of narration and evocation, can represent a long chain of actions very rapidly. Sensorimotor adaptations are limited to immediate space and time, whereas language enables thought to range over vast stretches of time and space, liberating it from the immediate. Whereas the sensorimotor intelligence proceeds by means of successive acts, step by step, thought, particularly through language, can represent simultaneously all the elements of an organized structure" (Piaget & Inhelder, 1969, p. 505).

Although verbal behavior means that children begin to express themselves and communicate, it is actually a unique form of communication. Initial verbal behavior of the preoperational child involves communication with oneself or what Piaget calls *egocentric speech*. Egocentric speech takes place during the ages of 2 to approximately 5, the preschool years. Later preoperational verbal behavior is characterized as socialized speech and involves communication with others.

EGOCENTRIC SPEECH

Preschool teachers should be aware of the characteristics of egocentric speech and adapt their instructions accordingly. Such verbal behavior cannot be changed or altered, and must be recognized as a natural part of the child's intellectual development and as something that will pass as the child gets older.

Essentially, egocentric speech can be characterized as a dialogue that children have with themselves. It is usually heard during playtime, which, at that age, occupies a large part of the child's time. Early preoperational children make no attempt to communicate with their peers. The following dialogue, recorded by Piaget (1977), is an example of egocentric speech of early preoperational children at play:

Mlle. L. tells a group of children that owls cannot see by day.

Lev: "Well, I know quite well that it can't."

Lev (at a table where a group is at work): "I've already done 'moon' so I'll have to change it."

Lev picks up some barley-sugar crumbs. "I say, I've got a lovely pile of eyeglasses."

Lev: "I say, I've got a gun to kill him with. I say, I am the captain on horseback. I say, I've got a horse and a gun as well." (p. 75)

Piaget describes this type of egocentric speech as a dual monologue, in which "the phrase recalls the paradox of those conversations between children . . . when an outsider is always associated with the action or thought of the moment, but is expected neither to attend nor to understand" (p. 70). Although it seems that Lev is communicating with someone, the someone is actually himself. There are two other types of egocentric speech, namely, *repetition* and the *monologue*. With repetition the child repeats words and syllables "with no thought of talking to anyone, nor even at times of saying words that will make sense" and in the monologue "the child talks to himself as though he were thinking aloud. He does not address anyone" (p. 70).

Around the age of 6 or 7 children's verbal behavior becomes more social, as indicated by an interest and concern in what their friends are doing and saying. Although socialized speech can not be equated with the intellectual "give and take" that adults experience, it does include an exchange of ideas in the form of critical remarks, commands, requests, and threats.

Such cooperative behavior of the later preoperational period can be seen in children's games where "following the rules" has a lot to do with winning and losing. This is one reason why preschool games should be less structured and more open-ended than games for 6- or 7-year-olds. Preschoolers are not very much concerned with following rules and cooperating with their peers to win. At first glance it will appear that there is some concerted effort to follow the game rules, but eventually the child will "play either by himself without bothering to find playfellows, or with others, but without trying to win, and therefore without attempting to unify the different ways of playing. In other words, children of this stage, even when they are playing together, play each one 'on his own' (everyone can win at once) and without regard for any codification of rules" (Piaget, 1977, p. 163).

We will now take an in-depth look at the actual thinking abilities of preschool children.

Chapter 3

INHIBITING CONDITIONS TO LOGICAL DEVELOPMENT

Introduction

The appearance of language heralds the beginning of the preoperational period and marks a profound change in the thinking ability of children whose intellectual development was previously limited to actions upon objects. The emergence of language allows the thought of the preoperational child to soar above his physical world. However, such soaring is limited!

Preoperational children can only think in terms of how they perceive things. This limited thinking affects their reasoning ability, which, in turn, makes it very difficult for them to understand such abstract operations as addition and subtraction. For this reason, these concepts should not be introduced until children are about 6- or 7-years-old, when their thinking is nearing the concrete operational level—Piaget's third stage of intellectual development.

There are at least six distinct but related factors which limit the preoperational child's ability to reason and think logically. They are:

1. egocentrism
2. states
3. centrations
4. irreversibility
5. transductive reasoning
6. nonconservation ability.

The preschool teacher should be familiar with each one and recognize them as part of the child's natural intellectual development.

EGOCENTRISM

Egocentrism, although somewhat similar to egocentric speech, has a wider application. Whereas egocentric speech mainly applies to the child's verbal behavior, egocentrism applies to the child's behavior in general: verbal and nonverbal. Consequently, it plays a definite role in affecting the child's overall thinking ability.

Egocentric behavior basically means that the child's thinking is self-centered. This type of behavior is different from the egocentric behavior of adults, however, in that, although such consistent behavior among adults is usually the exception rather than the rule, for preoperational children it is the other way around—these children believe that everyone thinks the way they do. They consider this thinking as the *right* thinking and, consequently, never feel that they have to answer to themselves.

According to Piaget (1968), egocentric behavior of the preoperational period has its origin in the sensorimotor period when "with respect to material objects or bodies, the infant started with an egocentric attitude, in which the incorporation into his own activity prevailed over accommodation (remodification of behavior as a result of experience)" (p. 18). To some extent, according to Wadsworth (1971), egocentric behavior is exhibited in all stages of development:

> Egocentrism is a characteristic of thought that is always present during the initial attainment and use of any new cognitive structure (schema). Later, it will be seen that the adolescent, upon developing completely logical thought, is very egocentric in his use of the newly acquired structures. Like other cognitive characteristics, egocentrism is not at constant level throughout the period. The child from two to four is much more consistently egocentric than the child from six to seven. As development proceeds, egocentrism slowly wanes until it is revived when new cognitive structures are attained. (p. 72)

The expression "you can't reason with a child" should be an important bit of advice for every preschool teacher who must recognize the preschooler's inability, as a result of egocentric behavior, to engage in any type of formal reasoning. This behavior inhibits cognitive development simply because the egocentric child "is untutored in the art of entering into other people's points. For him, there is only one comprehensible point of view—his own" (Gruber & Voneche, 1977, p. 115).

The existence of egocentric thought does not mean that preschool children cannot learn anything. It does imply, however, that they learn best with such less abstract and less formal 'hands-on' materials as blocks, rods, beads, and string.

STATES

Piaget also describes the logical thought of the preoperational child as tending to focus on states. This means that, at this stage, the child is unable to follow transformations in a sequence, but focuses only on states or certain aspects of a sequence. For example, a teacher holds a crayon in a vertical position on a desk. She releases the crayon and it falls onto the desk in a horizontal position. When asked to draw the falling

stages of the crayon, the preoperational child tends to draw either the vertical or the horizontal states of the crayon, but fails to conceptualize the falling crayon as undergoing a series of different steps or transformations.

Another factor inhibiting the logical development of the preschooler is a unique type of thinking called *transductive reasoning*.

TRANSDUCTIVE REASONING

Transductive reasoning is generally characterized as the child's inability to think inductively or deductively. In the formal operational period—the last stage of intellectual development—adolescents are usually capable of drawing conclusions from observable facts. They are also capable, when given some general rules, of deducing specific facts. The former type of reasoning is called *inductive* and the latter is called *deductive*. The preoperational child, on the other hand, can only reason from particular to particular or transductively.

The thought of the preoperational child "moves from particular to particular by means of a reasoning process which never bears the character of logical necessity. The child seeks neither to establish . . . a proposition by means of successive inductions, nor to postulate it for the purposes of deduction. Moreover, if we try to make him aware of a general rule, we shall find that it is by no means the rule for which we were prepared" (Gruber & Voneche, 1977, p. 106). For example, tell a preoperational child that all living things with four legs are called animals and that a lion is a living thing with four legs. Then ask for some more information about a lion. In all probability the child will describe other specific features of the lion, such as his whiskers or his roar, without making the conclusion that the lion is also an animal. His restricted reasoning capacity limits him to dealing only with specific aspects of the lion. He is unable to reason from a given general statement about animals to a specific fact about the lion.

CENTRATION

Preoperational thinking is also limited by how things look and appear. Centration is another example of the perceptually oriented type thinking of the child.

Some of us tend to read a book based upon its exterior appeal. That is, we "judge a book by its cover." This type of thinking typifies the preoperational child whose cognition is very much dominated by external characteristics of an object. These children tend to focus on specific traits or features of objects rather than to take into account all the features of the objects. Consequently, they are unable to decenter, that is, to go beyond the perceptual features of objects and to look upon them as representing a whole.

Evidence of preoperational children's inability to decenter can be seen by presenting them with two rows of like objects such as raisins. One row should contain eight raisins and the other six raisins. The row with eight raisins should be shorter than the row with six raisins as in the picture on page 30.

The child is then asked to choose the row that he would prefer to eat. In most instances the children will choose the spread out row (in our picture the bottom row), because to them, it appears to have more, and they are unable to rationally stand back and weigh the situation from the perspective of amount.

Some of the features inherent in centration were also seen operating in the children's inability to transform. They will also be seen in the next two topics: their inability to reverse their thinking as well as to conserve length and number.

IRREVERSIBILITY

The ideas of *irreversibility* and *reversibility* are very important concepts for Piaget. They essentially mark the difference in thinking between the preoperational and the concrete-operational child.

Preoperational children cannot reverse their thinking because of their fixation upon the perceptual aspects of their objects. Concrete-operational children, on the other hand, are capable of reversing their thinking and are consequently not limited in their reasoning by the appearances of things.

According to Wadsworth (1971), children who employ reversibility in their thinking are capable of following a "line of reasoning back to where it started" (p. 75). Preoperational children cannot follow a line of argument back to its beginning point. Their thinking is irreversible.

> The preoperational child's inability to reverse his thinking also has its origins in the sensorimotor stage. Preoperational thought retains much of the rigidity of sensorimotor thought even while surpassing it in quality. It is slow, plodding, inflexible, and dominated by perceptions. As such it remains irreversible. The attain-

ment of reversible operations is extremely difficult for the child. This is reasonable if one considers that all sensorimotor operations are irreversible by definition. Once a motor act is committed, it cannot be reversed. (p. 76)

One example of irreversible thinking occurs when the preoperational child pours water from a test tube into a beaker. She firmly believes that the amount of water in the beaker is less than the amount of water that was in the test tube.

The child is unable to realize that the quantity of water is the same in both containers regardless of their shapes. Her centering on a particular perceptual aspect of the experiment, namely the length of the test tube, inhibits her ability to think logically.

Further examples of the preschooler's inability to reverse thinking are evident in famous conservation experiments.

Non-Conservation Ability

There are several different types of conservation experiments. They range in complexity from one in length (usually given to ages 6 and 7) to the task in volume (given to ages 11 and 12). Preoperational children are not developmentally ready to respond correctly to these tasks because they cannot reverse their thought processes nor are they able to decenter. However, as egocentric thinking begins to wane, toward the end of the preoperational stage, the child is able "to decenter more and attend to simple transformations. All this in turn makes thought more reversible" (Wadsworth, 1971, p. 76).

Conservation tasks dealing with length and number are the standard experiments given to determine whether a child is starting to think concrete operationally.

In the "conservation of length" task a child is presented with two strips of paper of equivalent length. She is asked to find out for herself if these strips are of the same length. She usually does this by placing them next to each other and com-

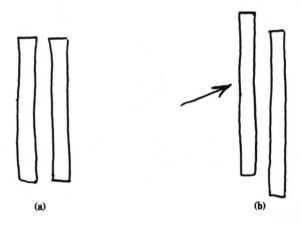

(a)   (b)

paring the lengths (a). The experimenter then moves one strip slightly ahead of the other and asks the child if both strips are now of the same length (b). The preoperational thinker invariably says "no" and points, after being asked, to what appears to her as the "longer" piece.

In the "conservation of number" problem, two rows of identical objects (here large white buttons), each row having the same number, are presented to the child (a). After the child

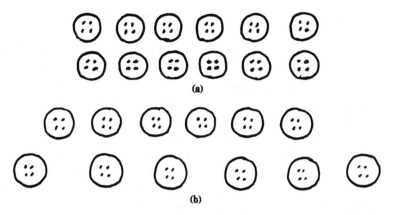

determines by a one-to-one comparison that each row has the same number of buttons, the lower row is spread out **(b)**. The child is then asked if both rows have the same number. Once again preoperational thinkers would point to the spread-out row as having more.

Children doing conservation tasks are as Wadsworth points out, "asked to judge which has more, is longer, etc. The change is always in an irrelevant dimension, though a correspondence remains. The object is to see whether the child can hold constant the quality being considered (conserved) in the face of the transformation. This requires that the child recognize a correspondence between the original form and the transformed one" (1971, p. 18).

An important result of these conservation tasks for the teacher deals with the idea of number. A preschooler has not developed a formal understanding of number and the reversible operation involved in such an understanding. Conservation tasks are difficult for them at this point of development "since the child has not yet acquired the notion of number, but only of perceptual wholes" (Gruber & Voneche, 1977, p. 324). Attaining an idea of number is an essential foundation for understanding an operation such as addition and its reverse—subtraction.

RESEARCH ON TEACHING CONSERVATION TASKS

Can preschoolers be taught how to conserve length and number so as to prepare them to start formal mathematics at an earlier age? Piaget believes that the child's ability to conserve should come naturally and should not be rushed by teaching and that only after the child has passed through the obstacles to logical thinking at a normal pace will he be ready to start conserving length and number. Thus, as Gruen has observed (1965), teaching conservation experiments to preschool children has no real affect on their later mathematical ability nor on their ability to remember them at a later time.

Chapter 4

CLASSROOM IMPLICATIONS OF PIAGET'S THEORY

INTRODUCTION

The ability of preoperational children to think logically is rather limited. Although the appearance of language allows them to express their thoughts verbally, their thinking ability is restricted by several factors. Their inability to take the other person's point of view or to go beyond the perceptual features of an object slows their reasoning abilities. As a result, their thinking is characterized as perceptual rather than conceptual. Not until they are 6- or 7-years-old can they go beyond the perceptual and begin to reason on their own without dependence upon how things look and appear. How then can we go about educating the preschooler?

As a cognitive psychologist, Piaget does not direct himself to the task of developing teaching methods and curricula for the classroom. He leaves that job for the curriculum specialist and educational psychologist. However, from his theories and his sporadic comments with regard to education, certain teaching methods and curricula can be inferred and made applicable to the preschool classroom.

SETTING THE FOUNDATIONS

Our new knowledge of how preschool children think should help us in arriving at some general strategies for teaching, and for devising curricula for them. By necessity their curricula will be different from those employed by the elementary school teacher.

Preschool educators have been criticized for their "incredible preoccupation with the teaching of language without coming to grips with how the preoperational child really thinks" (Kamii, 1973, p. 201). This does not mean that language development should be eliminated from the classroom. Having the child discriminate between "tall" and "short" objects or what "inside" and "outside" mean are good prerequisites for later math learning. However, the preschool teacher should be aware that, according to Piaget, thinking ability develops before language ability.

Teaching that emphasizes language development over thinking development is actually stressing memory over understanding. Too much "emphasis on language as a prime medium for thinking is bound to result in low-level activities that do not nourish intellectual development" (Furth & Wachs, 1974, p. 21). This means that preschoolers must be given the opportunity to experience new ideas before they learn them by rote— thus, for example, they should understand counting numbers instead of simply memorizing them.

Preschool teachers are limited in the extent to which they can develop the intellectual capacity of the child, for their inability to take the other's point of view makes it a trying task to impose upon them a set of rules for learning. As Furth and

Wachs point out, phrases such as "let's pretend" and "make believe" are more effective in capturing and holding their attention:

> The three to four-year-old child is easily fascinated by simple events in everyday life but is not ready for reflective explanations; he is interested in representative play rather than in the rules that underlie the play; he is intent on building a pretty house of blocks and not on a conscious reflection on the mathematical relationships implicit in the various sizes and shapes of the blocks; he is more likely to play alongside friends than plan and cooperate with them in a common task. (p. 273)

As children reach the age of 5, the change in their intellectual behavior makes them more amendable to following rules.

In shaping a curriculum for young children, their inability to decenter and to reverse thinking must be kept in mind—the preschool program should form a bridge between what they currently perceive as real to what they will later reason as real, and should contain the topics that form an adequate grounding for latter operational thought:

> Education has of necessity to be preparatory in the sense of familiarizing the child with the subject matter of the rules he will be learning later. Learning numbers and letters, drawing, practice in classifying and ordering materials, all acquaint the child with the material that he will order according to explicit rules at some later time. (Elkind, 1971, p. 50)

Therefore, there are two clearly related facts, Elkind continues, that teachers, who are about to select a teaching method and devise a preschool curriculum, should know about their student:

1. their language is not coordinated with their thinking;
2. they do not think or act according to rules as such. (p. 52)

A preschool curriculum that can adapt its educational program to include these general ideas will be in tune with the developmental age of the child; in general, this can be done by having it emphasize understanding rather than rote learning. It should also consist of simple-to-follow games and activities that will prevent preschoolers from mentally wandering off into their own "little world."

"TOOLS" IN THE PRESCHOOL CURRICULA

The fact that preschoolers are not capable of coordinating their thinking ability with their verbal behavior provides the teacher with the "natural starting point for educational intervention" (Elkind, 1971, p. 55). Every day, preschoolers face new ideas from such sources as the TV, parents, older brothers and sisters, teachers, and friends; yet only a few of these ideas will be of any real meaning to them. Most will be forgotten or used in a non-meaningful manner. For example, a 3-year-old might use the word "cat" to refer to what is in fact a squirrel. On seeing the squirrel for the first time, he easily mistakes it for a cat which he has seen many times before. Only when he experiences the difference between the two animals will he be able to intellectually digest the fact that one is called "cat" and the other a "squirrel."

In order to help "correct" this situation, the preschool curriculum must consist of "tools" that will enable children to experience ideas for themselves. This means that it should be action-oriented and include hands-on materials or what are usually referred to as *manipulatives,* such as toys, beads, buttons, etc. Once again, such a curriculum is based on the fact that a child understands (thinks) by experiencing and doing rather than by listening to words (language).

Although the manipulative type of curriculum should exist to some degree at all educational levels, it is especially appropriate for preschoolers who literally learn by doing. The experimential-based orientation gives language its proper guide and direction for as Piaget showed, doing and experiencing, rather than telling and listening are the keys to intellectual development:

> It would be a great mistake, particularly in mathematical education, to neglect the role of actions and always to remain on the level of language. Particularly with young pupils, activity with objects is indispensable to the comprehension of arithmetical as well as geometrical relations. (Gruber & Voneche, 1977, p. 727)

"To know an object" Piaget felt, "is to act upon it and to transform it . . . to know is, therefore, to assimilate reality in structures of transformation, and these are the structures that intelligence constructs as a direct extension of our actions" (Piaget, 1970, pp. 28–29). It is also for this reason that paper and pencil learning as well as workbook learning are the least effective learning modes for the preschooler. Both are removed from the concreteness of the manipulative level because a child cannot really act upon them. The use of workbooks and worksheets in the preschool should be minimal and limited to cases where, perhaps, some reinforcement of learning is desirable.

Certain concepts involving high degrees of abstraction take more time to develop than do simple concepts or ideas. For example, it . . .

> takes about six years of the child's life before he begins to have a mature class concept. It is abstracted from the child's actions toward objects, actions like comparing, contrasting, adding, taking away, sorting, observing similarities and differences, and evaluating what is an essential or an unessential attribute. What is abstracted here for intellectual growth is not from properties that are in the objects but from the general coordination of actions. (Furth & Wachs, 1974, p. 16)

The educational implication of Piaget's stance on language skills is then obvious. The stimulation of intelligent behavior by means of action or manipulatives should be foremost in the mind of the preschool teacher. Language development will come about as a natural byproduct of this "doing" process.

THE "SESAME STREET" SYNDROME OF EDUCATION

It must be noted, however, that some preschools differ in their approach to education. One such preschool curriculum is the television program, "Sesame Street." Although this program is probably one of the best on TV, it does have definite shortcomings in its attempt to educate due to the fact that "television promotes only 'figurative knowledge,' which is the static reproduction of images and perceptions based on memory or imitation. Television, therefore, cannot supply the means for producing 'true knowledge,' which, Piaget says, results from active manipulation of the environment" (Beck, 1977, pp. 16–19).

The main problem with preschool programs such as "Sesame Street" is that there is no way for children to actually touch and experience the differences in the sizes or shapes of the objects seen. Objects are shown on the screen, and the child is told about the differences between them. Any curriculum that teaches children by means of pictures (whether they be from a workbook or on TV) only, and allows no room for active manipulation of everyday objects, is not meeting the developmental needs of the preschool age child, according to Piaget. Learning of this type, if always used, is low-level, as its implicit educational philosophy actually rests on the memorization inherent in language learning.

In curricula of the "Sesame Street" type, says Beck, "the stated objective is to teach children how to think. But Piaget says children cannot be taught to think; they must learn for themselves by doing" (p. 18). The problem is that the writers of these curricula are not aware of the fact that children are capable of action and thought before they are capable of language, and so they deemphasize thinking in favor of language.

Developing a Teaching Strategy

Assuming that a preschool program adopts a manipulative, action-oriented curriculum, how does a child learn? How does a teacher go about instructing the child? Is the teaching formal or informal? It becomes obvious that preschoolers cannot learn in a structured formal atmosphere where the teacher occupies the front of the class and dispenses verbal knowledge. The egocentric nature of the preschooler will also prevent the child from learning in any school atmosphere where rules are enforced. The method of "teaching" a preschooler must, therefore, be somewhat different from teaching elementary-grade children who are of course, more capable of following another's point of view and instruction. Teaching the preschooler is another matter, Elkind points out, for he "does not see any necessity for acquiring new information since he believes it is already stored in his head. Formal or direct education involving the inculcation of facts and rules is inappropriate for such a child because he does not accept the underlying contract in such an arrangement; namely, that he has something to learn from the teacher" (p. 52).

On the other hand, it would not be practical or sensible to create a teaching atmosphere where the preschooler is allowed to do anything he pleases. An unstructured, totally open-ended environment will lead ultimately to boredom for teachers and students. Consequently, the best teaching approach is one that is somewhere between a structured and nonstructured one, with an eye also at determining the individual needs of each student.

Although preoperational children tend to exhibit the same thinking characteristics at some point in time, not all of them develop at the same rate. Some preschoolers, because of their background and experiences, will need more guidance and structure in their educational program than other students. Others might have to be referred to a learning specialist to diagnose a possible visual-perceptual or behavior problem. On the other hand, some will be capable of structuring their learning environment by themselves with little guidance from teachers. The job of the preschool teacher, says Kamii, is to determine "what the learner already knows and how he reasons in order to ask the right question at the right time so that the learner can build his own knowledge" (p. 203). Once the needs of the child are determined by the teacher "her function is to help the child construct his own knowledge by guiding his experience" (p. 212).

"Guiding his experiences" does not mean, of course, that the preschool teacher should tell students how to do things. We know that the developmental nature of the child dictates that he is incapable of sitting back and taking constant verbal instructions from a teacher. He must be given the freedom to experience and interact with his surroundings. Perhaps, then, one of the best teaching strategies for the preschooler is the guided discovery technique. This technique is not unique to preschoolers. It could be employed very well in the elementary and higher grades when a lesson warrants its use.

GUIDED DISCOVERY LEARNING

It is very difficult to give a precise definition of learning by discovery since teachers use it in a variety of ways. Basically the discovery technique places emphasis upon the preschoolers' ability to "discover" ideas themselves as a result of their interaction with the environment. Piaget himself describes such an idea:

> it seems that the subject has been able to discover for himself the true reasons involved in the understanding of a situation and, therefore, has at least partially reinvented it for himself. (p. 731)

The discovery technique changes the role of the teacher who is now "less that of a person who gives 'lessons' and is rather that of someone who organizes situations that will support such behavior by means of appropriate arrangements. Should the child have difficulties in his attempts to grasp a certain idea, the procedure with an active methodology would not be directly to correct him, but to suggest such counter examples that the child's new exploration will lead him to correct himself" (p. 731).

There is no one way that will lead a student to discovery. In his article, "On learning Mathematics," Jerome Bruner offers several suggestions:

1. a Socratic method (the asking of a series of easily answerable questions leading to a logical conclusion to be arrived at by the student);
2. making up certain sequences of pattern-type problems that allow a student to find regularities;
3. the projection by the teacher of an attitude of interest, daring, and excitement;
4. providing the student interesting clues to short cuts so that he may discover for himself other techniques for short cuts. (p. 610–619)

The following dialogue is an example of using the Socratic method when discussing with a 5-year-old the concept of the rational number ten:

> *Teacher:* How many fingers are on my hands?
> *Student:* I don't know.
> *Teacher:* Don't you remember what we did with each of our hands yesterday?
> *Student:* What?
> *Teacher:* I'll hold up my two hands and wiggle my fingers. Do you remember now?
> *Student:* Oh, yes! We counted them. One, two, three, . . .
> *Teacher:* Now, how many fingers do you have on your hands?
> *Student:* Hmmm??

Teacher:	What did you just do with my fingers on my hands that you can do with your hands?
Student:	I can count them, too. One, two, three, ...
Teacher:	Now, how many fingers do you have?
Student:	Ten, too!
Teacher:	Do you see all those red blocks on that table? Do you think you can bring me back a pile of just ten red blocks?
Student:	(Brings back a pile of blocks and then counts ten out).
Teacher:	Is there anything else in the room that you can use to make a set of ten?

Note that in the dialogue, the role of the teacher is to guide the student to an understanding of ten by asking questions that she can easily answer. Through this mode of questioning the student should logically come upon the knowledge of the concept of ten by herself rather than being told by the teacher what ten means.

Some topics in the preschool math curricula lend themselves readily to the second suggestion for bringing about discovery. The teacher can ask the student to find the blocks that will complete the following shape pattern:

In an ordering lesson she can ask the 5-year-old child to try to find the next stick:

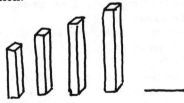

A lesson on rational counting can require the student to find the correct "number" and place it above the appropriate numeral:

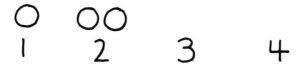

It is not too difficult to instill an "attitude of interest, daring, and excitement" into the lesson. Preschool children are at an age when a diversified number of ideas are being experienced for the first time. With appropriate comments and remarks, it is not very difficult for the teacher, for example, to get her students excited and interested about making a mobile to take home to mommy and daddy, going on a trip to a local farm, or playing a counting game.

The teacher can also provide the students with "interesting clues to short cuts" by giving them general hints instead of direct answers to their questions or problems. General comments such as "rearrange the blocks differently and now tell me what you see" or "do it this way instead and let me know what happens" might allow children ways of seeing and doing problems that they were not able to see before or perhaps to solve them now more quickly.

The basic tenet behind discovery learning is for the student to become actively involved in the learning situations with the appropriate guidance from the teacher. Since preschool classrooms are usually arranged in an informal manner where individual and small-group work can take place, this teaching technique is probably the most effective. Discovery teaching coincides well with a well-known expression whose last line should typify the learning environment of the preschooler:

I hear and I forget
I see and I remember
I do and I understand

Chapter 5

A LOOK AT OTHER
TEACHING-LEARNING TECHNIQUES

IINTRODUCTION

The guided discovery technique has its limitations. It probably will not work well with preschoolers who come from disadvantaged backgrounds or who have a learning or an emotional problem. These children need attention from their teachers in the form of a more structured learning environment. Using the Socratic teaching technique on them will probably lead to very little self-discovery and a lot of frustration for teacher and student.

Even some children reared successfully on the guided discovery technique will require a more structured approach when experiencing a new or challenging topic. Although the guided discovery technique is perhaps the best all-around teaching method for the preschooler, it evidently cannot be used all the time in the classroom. The purpose of this chapter is to provide the preschool teacher with teaching-learning techniques to use when it is impractical to use only the guided discovery method.

TEACHING BY GAGNE'S HIERARCHY

Robert Gagne is a behaviorist whose theory about how people learn differs from the cognitive (intellectual development) approach of Piaget. With regard to learning, Gagne holds that "external organization is a necessary condition to optimize learning. The cognitive position assumes organization lies within the individual" (Wadsworth, 1971, p. 126). In terms of learning, "external conditioning" is a key idea for Gagne and "internal development" is the key idea for Piaget. However, it is suggested that they do have something in common since both of them "conceptualize learning (concept acquisition) as taking place in a manner that is orderly, sequential, integrative and hierarchical." (p. 124)

Regardless of their theoretical differences, teachers tend to be pragmatically oriented in their classroom techniques. If they find a teaching method, such as Gagne's, that works well in a particular case, they will probably adapt it.

Gagne outlines a learning sequence involving six basic steps that can be used as a guide by any teacher or curriculum writer for helping a student through a specific learning task. These steps are sequentially arranged from simple to complex:

1. CHAINS.   Chains are the first of the six steps in the learning process. Generally, according to Gagne, a chain is "a sequence of individual responses arranged in such a way that the entire set of responses reels itself off from start to finish" (1971). There are two types of chains: motor and verbal.

Motor chains "include actions such as a linking together of all the individual responses necessary to hold a pencil, turn on a light, cut paper with scissors, kick a soccer ball, or ride a bicycle," while "verbal chains include the memorized sequences of words" (Galloway, 1976, p. 98). Verbal chains may be either "extremely simple, such as 'come here,' 'help me,' or 'I don't want to,' or they may be extremely complex, such as in the rote memorization of the letters of the alphabet, definitions for words, or perhaps all the words of a long poem" (p. 98).

Chains form a basic foundation upon which the remaining steps in the learning hierarchy rest. It is important to realize that they do not involve any rational understanding on the part of the student. According to Gagne, a 4-year-old might be able to count by rote (verbal chains-complex) from one to ten without any rational understanding of what each number actually means. Because Piaget, as we have seen, does not believe in rote or verbal learning, we will concentrate on those aspects of chains that do not involve such memorization.

2. DISCRIMINATIONS.   In discriminations, according to Galloway, a learner should be able "to tell whether two things are the same or different . . . not that the learner can tell what the stimulus objects are or be able to name them" (p. 98). For example, is the preschooler able to see that a set containing two blocks is different from a set containing three blocks without mentioning the difference?

Or, from a set of blocks containing math triangles and circles, can she sort the different shapes into separate piles:

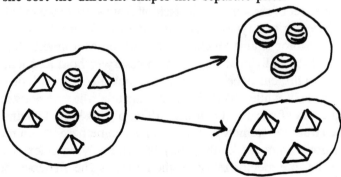

3. CONCRETE CONCEPTS.   Recognizing what makes objects differ from each other comprises the next step in Gagne's learning schema. Concrete concepts are "learned when the individual can recognize or identify an object quality, like round or square, or an object like a chair, by its appearance" (p. 98). The child is now asked to describe the physical properties and characteristics of the objects that were previously classified as being the same or different—for example, one might be round, the other square; or one might be blue, the other red; or one might be large, the other small.

4. DEFINED CONCEPTS.   Defined concepts are learned when the preschooler can identify a concept by using a definition. In the case of a 3-year-old this can be seen when she points to the set containing three objects:

5. RULES.   This process involves understanding a relationship between one or more concepts. For example, preschoolers would probably understand the rule that states that two triangular-shaped blocks can be formed into a rectangular shape:

6. HIGHER-ORDER RULES.   These are "composed of two or more simpler rules. Often a learner puts together simpler rules into higher-order rules in a problem-solving situation that is new to him. He discovers the higher-order rule" (p. 98). For example, can a preschooler apply the rule mentioned above and

find objects that are made of two triangles, such as a door, a book, or a box? Notice that this step involves some discovery on the part of the child.

Because Gagne's method involves a linear ordered model in which a child cannot proceed to the next step unless he has done the previous step, his hierarchy offers some structured help for those who need it. However, not all preschool topics can be taken from chains to higher-order rules.

If a topic can be put through all the steps, these would be the general questions asked for each:

1. *Chains.*   Can a student hear and then follow simple directions? Can he go through simple motor skills without too much difficulty?
2. *Discriminations.*   Can he see, taste, smell, hear, and feel the sameness or difference in the qualities of objects?
3. *Concrete Concepts.*   Can she tell if objects are round, tall, flat, smooth, furry, rough, blue, green, etc.?
4. *Defined Concepts.*   Can he point to objects by name such as a "chair," the number "3," a "circle," a "blue car," etc.?
5. *Rules.*   Is she able to understand a relationship between two concepts, such as "yellow *mixed* with red produces orange" and "a set with three circles has the *same amount* as a set with three triangles?"
6. *Higher-Order Rules.*   Can a student use rules in new situations such as discovering other colors that will produce a new color, different ways of forming sets of three, etc.?

Going through at least the first five steps with the child will enable the teacher to determine the exact nature of the student's problem, if one happens to exist. A child who might be experiencing some difficulties in seeing (discrimination) the difference between two objects might be having a visual or a visual-perception problem and should be referred to a learning or medical specialist. A preschooler who will not follow simple directions could be suffering from an emotional or behaviorial disorder. Or, if a preschooler fails to understand a set containing five elements he might require more basic prerequisite work. Figures 5.1, 5.2, and 5.3 show ways of using the hierarchy on various topics in the preschool curriculum.

Gagne's hierarchy provides the preschool teacher with an efficient and systematic teaching tool for students, who need it.

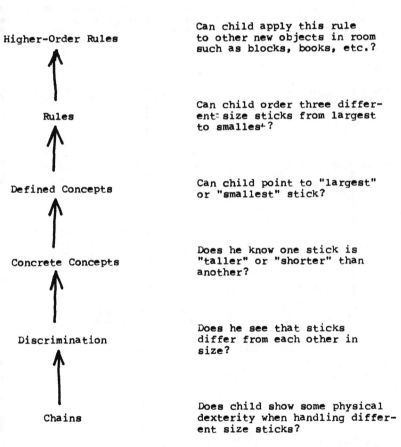

Figure 5.1   Ordering Three Sticks of Slightly Varying Size

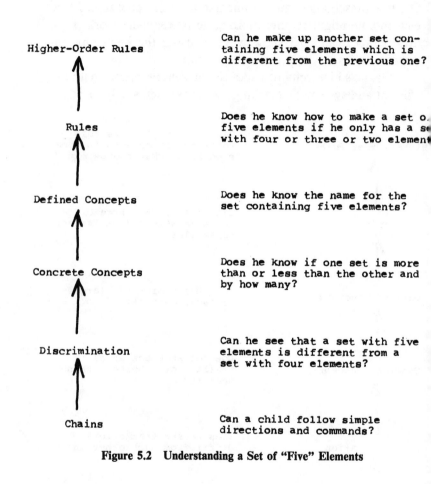

Figure 5.2   Understanding a Set of "Five" Elements

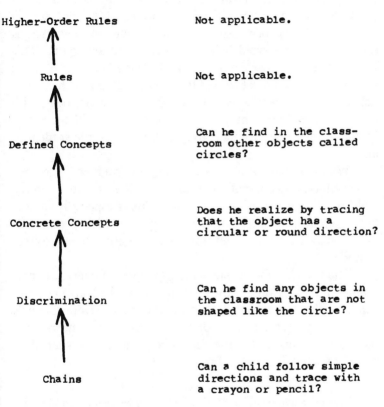

Higher-Order Rules	Not applicable.
Rules	Not applicable.
Defined Concepts	Can he find in the class-room other objects called circles?
Concrete Concepts	Does he realize by tracing that the object has a circular or round direction?
Discrimination	Can he find any objects in the classroom that are not shaped like the circle?
Chains	Can a child follow simple directions and trace with a crayon or pencil?

Figure 5.3   Understanding the Properties of a Circle

53

Individualization and Skinner's Operant Conditioning

Another behaviorist whose theories are helpful in attempting to solve certain learning problems is B. F. Skinner. Skinner's theory, the principle of operant conditioning, simply states that living organisms tend to repeat behavior that is satisfying and to avoid behavior that is not satisfying. This implies that a teacher can shape the learning behavior of those pupils who would most benefit from such a change. This external conditioning is most successful when the teacher deals with only one preschooler at a time. In fact, as we will see shortly, the teacher does not have to actually be with the student at all.

While shaping a child's learning environment, the processes underlying operant conditioning also tend to instill in the child positive ideas toward learning. This is especially important in an area such as mathematics, about which young children too often tend to develop more negative than positive attitudes.

The teacher who employs the principles of operant conditioning must be aware of the exact nature of the learning problem and the steps needed to overcome them. Here Gagne's hierarchy can serve as a guide. Preschool math concepts can then be so arranged that the child's behavior can be shaped by means of what is called *reinforcement:*

> Operant conditioning says little about group behavior. Instead, it shapes individual behavior by starting with an individual operant shaped by reinforcements to each individual student, which are supplied according to an individual's responses or groups of his responses. (Higgins, 1973, p. 133)

Teaching, according to Skinner (1963), is then the "arrangement of contingencies of reinforcement under which students learn" (pp. 64–65). This arrangement is done through a

Valerie works on a "teaching machine."

"You're right!"

sequential positioning of material from easy to complex that is
often referred to as "programmed instruction" or "a teaching
machine."

In programmed instruction, the preschooler is told or reinforced immediately with regard to the correctness of his responses without the need of a teacher. Since the least complex
"teaching frames" are usually placed in the beginning of the
instructional unit, the student is given the opportunity to experience immediate feedback when his answers are correct. This
process instills a positive attitude toward learning.

The preschool teacher can instruct a 5-year-old in the use
of a "teaching machine" in a lesson dealing with matching
objects and numerals. Immediate reinforcement is obtained
when the "shapes" match up.

A teacher can also construct a device that will allow a child
to work alone on a lesson on labeling a set with the correct
numerals. Here the teacher only has to start the child by doing
the first step:

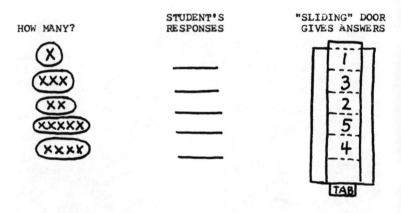

Still another teaching machine idea is an educational toy
that deals with the concept of ordering. It is called a "color
cone," and consists of about ten rings of various sizes. When
ordered correctly on a stick, the toy is shaped like an inverted
ice cream cone. The teacher can show a 5-year-old what the
final shape should be when the task is correctly completed:

Programmed materials for the preschooler will vary depending upon the resourcefulness and thinking of the teacher. Many of them can be made simply with oaktag, tape, and magic markers and should only be used when the child needs extra help or some reinforcement. Programmed materials should not be the primary means of instruction for the average preschooler who needs to use guided discovery techniques while exploring manipulatives. However, they do have their place in the preschool classroom and children enjoy learning from them:

> In arranging contingencies of reinforcement, machines do many of the things teachers do; in that sense, they teach. The resulting instruction is not impersonal, however. A machine presents a program designed by someone who knew what was to be taught and could prepare an appropriate series of contingencies. It is most effective if used by a teacher who knows the student, has followed his progress, and can adapt available machines and materials to his needs. Instrumentation simply makes it possible for programer and teacher to provide conditions which maximally expedite learning. (Skinner, 1963, pp. 168–177)

Chapter 6

CONCLUDING REMARKS

Psychological Rationale for Part ii Topics

Many of the activities presented in the second half of this book can be arranged under three major headings: *classification, ordering,* and *geometry.* These topics were chosen because they play a fundamental role in the development of each preschooler.

Piaget believes that classifying is a natural part of the child's psychological growth and begins very early in the sensorimotor years:

> As children explore the world in which they live, they learn to recognize and name the various objects they see. Among the first objects recognized, for example, are mother and the feeding bottle. Later, other objects are pointed out and named, such as, car, house, or dog. These objects are recognized on the basis of certain physical properties, such as, color, size, shape, or certain patterns of behavior. In being able to recognize an object, the child has classified it into a certain category different from many other objects, based on certain unique characteristics or properites. (Copeland, 1979, p. 62)

Similarly, ordering or seriation "is a very primitive ... structure in children's thinking, just as primitive as the classification structure" (Piaget, 1970, p. 28). At a very early age children are naturally capable of ordering three objects that differ to a degree in size:

As early as one year of age, or as the child is moving into the sensorimotor level, he can order by size three objects such as bricks if the size differences are easy to recognize, thus solving the problem on a perceptual level. (Copeland, 1979, p. 90)

Geometry is a branch of mathematics that deals with the position of points in space. These points are abstract entities and are usually represented by dots (.) on a paper. According to Piaget, the child's first concept of space is topological rather than Euclidean:

The child's first impression of space or the world in which he lives is a very disorganized one. Figures come and go before him as in a moving tableau. Shape to him is not a rigid thing. The shapes he sees are often changing ones. The door looks different as it is opened. His mother looks different as she moves closer. Her face (to him) changes in shape as it turns to right or left and down as it moves nearer or farther. (Copeland, 1979, p. 255)

This means topological terms such as *inside, between* and *outside* should be taught before the ideas of squares, triangles, and rectangles are presented to the preschooler:

Not until a considerable time after he has mastered the topological relationships does he develop notion of Euclidean . . . geometry. (Piaget, 1953, p. 75)

Since preoperational children are in various stages of intellectual development, the activities in the next half of this book will be broken down into different levels of difficulty. Early preoperational children are less capable of logical and abstract thought than later preoperational children, who are usually not very far from being able to conserve number and length.

SUMMARY

Part One is intended to serve as a background and introduction to Part Two, for teachers cannot adequately construct learning activities unless they know something about the intellectual abilities of their students. Particularly in the fast-growing area of preschool education, it is not unusual to find educators lost and confused about "how" and "what" to teach 3-, 4-, and 5-year-olds. Part One was an attempt to examine the "how" question.

Part One also stressed that parents must free themselves of the misconception that they are doing something wrong when they send their 3-, 4-, or 5-year-old to school. Most children thoroughly enjoy their preschool experiences as the parents will quickly realize when their child comes home eager to tell them "what I did at school."

Part One also reviewed Piaget's findings that thinking ability develops well before language ability in the child. Around 2 years of age, the child begins to use coherent words and in some cases says simple sentences. The new acquisition of language allows the thought process to develop at a very fast rate. The child is no longer limited in expression to pointing, facial gestures, or body movements. The thinking displayed by the preschooler, known as egocentric thinking, is unique to this stage of development, however, and is characterized by the inability to take the viewpoint of another. Accordingly, the teacher must learn to adjust his instruction.

Since children's verbal behavior is not yet in tune with their intellectual behavior, the discovery technique is more effective than the expository technique in helping them learn. By organizing children's learning environment, so that there is an emphasis on manipulatives, preschoolers learn to discover experiences by themselves. When the discovery method falls short of achieving its goal, the teacher should consider adapting the direct methods of Gagne and Skinner.

The activities in Part II can be incorporated directly into the classroom as they appear, or they can be adapted and modified as the teacher finds necessary. However they are used, it is hoped that both teacher and student share in the fun and excitement of doing them!

Part II

ACTIVITIES

Chapter 7

ORGANIZING THE PROGRAM

SHOULD MATHEMATICS BE INCLUDED IN THE PRESCHOOL PROGRAM?

Asked the question that heads this chapter, teachers' answers vary from a resolute, "No!" to an enthusiastic, "Of course." These are some familiar responses:

1. "Absolutely not. Math is too hard." This might be the right answer. Much of what comes under the heading of mathematics is too hard for preschoolers. Piaget cautions educators to introduce topics at the optimum moment. The child must be ready to learn and this readiness is dependent upon the development of innate logical thought. Topics introduced should be within the realm of the child's understanding, not too hard, but difficult enough to provide growth.
2. "No. Preschool is supposed to be fun." Yes and no. Yes, early learning should be enjoyable and rewarding.

It should make children feel good. But math can do that. The activities suggested in this book are developmental, they never pressure the learner, they do not require extensive drill.

3. "Yes. Children should learn to count." True, counting is one aim of the preschool math program but it is not the only aim. In the following pages are sequential activities leading to understandings that have often been left to chance.

4. "I suppose so. What do you have in mind?" Good answer. It shows a willingness to accept some suggestions. It has been stated by Piaget and others that the development of logical thought should not and in fact cannot be accelerated. However, he firmly believes that the proper environment including a *questioning* teacher will insure each child's development at the optimum rate. The lessons which follow are not "telling" lessons but are "questioning" lessons. They are not so simple as to be boring. Mostly, they are individual activities since each child is not expected to develop at the same rate. They provide clues so the alert teacher can determine each child's level and they suggest means of gently prodding his or her cognitive development.

5. "Yes. Young children use math skills in their everyday life. Besides, they enjoy it." Much better. Some children perceive and accomodate information from the world about them more readily than others. A kindergartener was asked how he knew a particular thing. He placed his hands on his hips and replied, "I've been alive for five years, haven't I?" He was a rapid learner. Others need much more guidance in abstracting knowledge from their environment. Asking questions about particular manipulative materials is one way to bring about awareness.

6.   "Yes. Math skills are essential in adult life. A developmental program based on children's abilities will provide an essential firm foundation for formal mathematics." The key word here is "developmental." The pages that follow contain activities to meet children's needs as their logical thought matures.

When Can Math Be Taught?

The school day is short. There are so many important things to include. When can math be squeezed into an already tight schedule? A determined teacher is the most important ingredient. If the teacher is convinced of the need to include math, if the children obviously enjoy the activities, time will be found each day.

Perhaps the following suggestions will help:

1.   Planned math lessons

   a.   Individual lessons. Ideally, each child should have the benefit of an adult's full attention at least once a week while working on lessons planned specifically for him or her. Some preschool teachers with classes of from 15 to 25 youngsters manage this by allowing each child a "turn" for 10 minutes during playtime. If more than one adult is available, the time can be increased. This arrangement makes certain that no child is overlooked.

   b.   Small groups. Perhaps an activity would be appropriate for two or three children, or maybe it would be more enjoyable to work with others.

   c.   Entire class. Some activities (story telling, flannelboard work, games, artwork) are too much fun to exclude any of the group. Plan two 15- to 20-minute class lessons per week.

2. Include math concepts in the routines of the day

   Are there enough chairs?

   Please get one straw for each child.

   Is it warm enough outside to go without our sweaters?

   Laurie has one day less to wait for her birthday.

   Are all of the children here today?

   How many are not here?

3. Integrating math with other subjects

   Bring math concepts into art, science, music, health, physical education, holidays. The possibilities are endless.

4. Incidental teaching

   There is no better way to convey the usefulness of mathematics in everyday life than to capitalize on events in the classroom. Tommy brings two rocks to school. After admiring his treasures, the teacher shows further interest with questions as, "Which one do you think is larger? Which is heavier? Would they fit into our fishtank? Do you think the water would overflow?" "Show and Tell" provides innumerable opportunities for meaningful math lessons.

WHAT ARE THE AIMS OF THE PRESCHOOL MATHEMATICS PROGRAM?

Each of the chapters which follow concentrates on a topic which is considered to be an important readiness concept for later mathematics. Within each chapter there is first an educational rationale of how the topic relates to the learning of formal mathematics and then activities which are arranged sequentially. Level I activities are appropriate for most 3-year-olds, level II for 4-year-olds, and level III for kindergarteners.

Chapter 8

CLASSIFICATION OR SORTING

Before the child can bring meaning to abstract mathematics he must come to terms with the concrete world about him. The 7-year-old who says, "You may have the red jelly beans and I will take the others," has made an observation, applied logical reasoning, and quickly ends up with the most jelly beans. She saw the set, noticed likenesses and differences, separated the set into two subsets (those that are red and those that are not red), compared the two on the basis of their numbers and made an advantageous decision. Since the process is more or less spontaneous, she probably would not be able to verbalize her chain of thought. While this thinking seems apparent to the older child, it is beyond the 3-year-old's cognitive development. He sees the candy and wants "some."

One of the first aims of the preschool math program is to help each child develop the ability, through the use of the senses, to impose an organization on a set, to notice that some things are alike in some way and therefore belong together. As cognition develops, the aims expand to keep pace with the increasing flexibility of thought.

The activities in this chapter give children the opportunity:
1. To sort a set based upon a common property;
2. To discover another similarity and to regroup accordingly;
3. To perform multiple sorting by finding likenesses and differences within a subset;
4. To sort according to two properties;
5. To classify on the basis of negation.

ACTIVITIES

What's Alike, What's Different?

Purpose:	To help children notice likenesses and differences
Level:	I
Number of players:	3 to 5

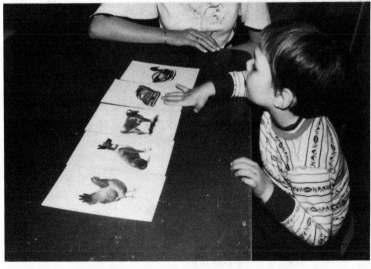

Five-year-old Michael has no trouble choosing the *different* one.

Materials: From around the room collect items which are alike in one way yet different in another, such as:
two blocks, same shape, different size
a crayon and a pencil
a coat and a sweater
Two beads, same color, different shape
toy truck and doll
banana and apple
pictures of a horse and a dog

Procedure: Show two of the items. Ask children how they are alike and how they are different.

Adaptation: Show two pairs of items. Ask children to choose the one from the second pair that belongs with the pair you are holding.

Feel for It

Purpose: To reinforce the meanings of words of comparison and position

Three-year-old Brett finds a circle shape in the feel box.

Level: I

Number of players: 1

Materials: "Feel Box"
Items from around the room such as:
a straw and a pencil
a ball and an inflated balloon
fake fur and sandpaper
two blocks of different sizes
a fork and a spoon
a hat and a mitten
a doll's shoe and a child's shoe

Procedure: Place two items inside the box. Ask, "Can you find the one that is:
the longer (or harder, or hollow)
the heavier (or larger, or softer)
rough (or feels best)
smaller

used to eat ice cream
worn on your head
too small for you

Which Doesn't Belong?

Purpose:	To help children notice differences
Level:	I
Number of players:	3 to 5
Materials:	Sets of three objects, one of which is different from the other two such as: two picture books, one coloring book two different shoes and a sneaker doll-house furniture - chair, sofa, and a table a rake, a hoe, and a hammer
Procedure:	Ask a child to pick out the one that doesn't belong with the other two

Picture Cards

Purpose:	To help children notice differences and likenesses
Level:	I
Number of players:	1
Materials:	Set of picture cards, approximately half showing two objects, that are identical, the rest showing two objects that differ in some way

Procedure:	Ask a child to put the cards into two piles, the *alike* pile and the *different* pile. For self checking, the backs of the cards can be colored.

Sorting Blocks

Purpose	Give experience in directed sorting; teaching colors or shapes
Level:	I
Number of players:	1
Materials:	Attribute blocks Three plastic bags
Procedure:	Say, "Blocks of each color have their own plastic bag. This block is red and it goes in this bag. Find all of the blocks that are the same color and put them in here." Repeat for blue and yellow.
Variation:	Have children sort by shape.

Larger, Smaller

Purpose:	Provide practice in directed sorting; teaching proper use of *longer, shorter*
Level:	I
Number of players:	1
Materials:	Kingergarten building blocks of same shape but of two sizes Two cardboard cartons
Procedure:	Direct the child to put all of the shorter blocks in one box and the larger ones in the other box.

Kitchen Helpers

Purpose: Provide experience in directed sorting
Level: I
Number of players: 1
Materials: Silverware
Drawer organizer with indentations shaped for the utensils (available in most variety stores)
Procedure: Briefly discuss the shapes of the indentations. Ask the child to sort accordingly.

What's Missing?

Purpose: Improve visual discrimination
Level: I or II
Number of players: 3 to 5
Materials: Set of picture cards, each showing the same object twice but with something missing in one of them.
Procedure: Have children take turns guessing what is missing. To reinforce this lesson children can draw the missing part on dittos.

Community Helpers

Purpose: To teach children to identify the tool used by each worker

Level:	II
Number of players:	3 to 5
Materials:	Pictures of men and women in various occupations: carpenter, farmer, doctor, plumber, etc. Children can help collect these from old magazines. Picture cards each depicting a tool used by one of the workers.
Procedure:	Have children take turns matching the tool to the worker.

What Is It For?

Purpose:	To teach children to identify items by use
Level:	II
Number of players:	1 to 4
Materials:	A set of picture cards of: things to ride in things to wear things to eat things that give light things to feed pets building tools garden tools
Procedure:	Have one child sort the cards into the appropriate piles. An adaptation of "Go Fish" can be played by two or more children. Four or five cards are dealt to each player. The children take turns asking for cards in a particular category.

Alphabet Blocks

Purpose:	Provide practice in directed sorting
Level:	II

Number of
players: 1 to 3
Materials: Alphabet blocks (blocks with letters, pic-
                  tures and numerals on the faces)
Procedure: Have children take turns tossing a block. As
                  each is played it is grouped with those of
                  similar top face. Three children can play,
                  each attempting to toss one category.

Buttons

Purposes: Directed sorting
                  Reinforcing colors
Level: II
Number of
players: 1 to 4
Materials: Many assorted buttons
                  Lengths of yarn threaded into blunt needles
Procedure: Have children string the buttons by color

Autumn Leaves

Purpose: Practice directed sorting
Level: II
Number of
players: Class
Materials: Many colored leaves collected during an au-
                  tumn nature walk
                  Bulletin board or poster paper
Procedure: Back in the classroom after the nature walk
                  ask the children to find the leaves that might
                  have grown on the same tree. Shape is the
                  only consideration. Children will have to
                  disregard color and size. Arrange a colorful
                  display.

Sorting Box Activities

Three-year-olds using a sorting box.

Purpose:	Practice directed sorting
Level:	II
Number of players:	1
Materials:	A sorting box. This is easily made by cutting slits into the lid of a shoe box. A strip of oaktag is folded to form the separators. Materials to be sorted: geometric shapes picture cards of things we eat, wear, play with picture cards of birds, fish mammals holiday pictures—things relating to Halloween, Thanksgiving, Valentine's Day (holiday stickers on index cards are quite satisfactory)

Procedure: Have child sort according to directions. After sorting, the child can remove the lid to check if the items in each section are alike.

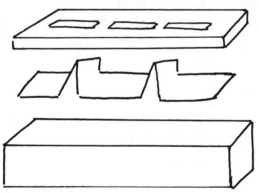

Undirected Sorting

Purpose: To develop the ability to notice similarities

Level: II

Number of players: 1

Materials: A box of beads and blocks
An empty box

Procedure: Ask the child to "straighten these out." If the child does not see the implication, return to directing the sorting.

Variations: 1.   Attribute blocks with directions: "Put the ones together that belong together."
2.   Collection of nails, screws, washers, say, "Fix these the way they belong."
3.   Box of assorted crayons.

Multiple Sorting

Purpose: To develop the ability to sort according to two properties.

Level: II

Number of players: 1 to 3

Materials: Attribute blocks

Procedure: First, ask the children to sort by color. Next, have each of these subjects sorted by shape (or size or thickness).

Guess What I Am

Purpose: To teach children to identify an object by the sound it makes.

Level: II or III

Number of players: 4 to 8

Materials: Picture cards of things that make identifiable sounds such as: a dog, a train, a cow, a baby, a fly. Children may enjoy finding their own pictures in old magazines.

Procedure: Have one child begin by choosing a card and holding it so the others cannot see it. He then makes the appropriate sound and chooses another player to guess the object. The first to name the object gets to be "it."

Which Doesn't Belong?

Purpose: To determine if a child is able to reverse his or her thinking and see a situation from another viewpoint.

Level: II or III

Number of players: 4 to 8

Materials: Sets of three objects with two possible groupings such as: a lead pencil, a red pencil, a red crayon; pictures of a dog, a goldfish, a squirrel; pictures of a sled, a wagon, an automobile; pictures of an ice cream cone, a sandwhich, a dish of ice cream; pictures of a fishing pole, a fish, a frog; Attribute blocks: small red triangle, small blue triangle, large red triangle; two baby dolls, boy and girl, one adult doll; cup, saucer, drinking glass; pictures of a child wearing glasses, adult wearing glasses, child without glasses.

Procedure: The dialogue might be as follows:

Teacher: Girls and boys, look at these things on the table. We are going to play a guessing game. I think that one of these doesn't go with the others. Who would like to pick up the one that doesn't belong?

Eric: I know which one. It's this. (He holds up the crayon.) Those are pencils and this is a crayon.

Teacher: That is a very good idea. Does anyone else have a different idea? (She returns the crayon to the table.)

Laurie: Maybe it's this one. (Chooses the lead pencil.)

Eric: That's not right. That's a pencil and this is a pencil and this is a crayon.

Jeff: I pick the crayon but Laurie could be right, too. The pencil she picked isn't red like these are.

Jeff's comment shows he has reached a flexibility of thought beyond that of Eric. The ability to reverse opinions which are based on sensory imagery shows a higher form of reasoning, according to Piaget. Eric is not being willfully stubborn, he simply is unable to accept a viewpoint other than his own. More docile children may appear to be "learning" by agreeing with the teacher's verbal explanation, but further questioning might reveal no real understanding.

Some or All?

Purpose:	To check on understanding of the terms *some* and *all.*
Level:	III
Number of players:	3 to 5
Materials:	4 red triangles
	1 blue triangle
	3 blue squares
	(Attribute blocks may be used or the figures could be cut from construction paper.)
Procedure:	After allowing children to view the figures and discuss them, ask, "Are all of the triangles red? Are all of the blue ones squares?" Don't be surprised if the responses are incorrect.

Which Has More?

Purpose:	To determine if a child comprehends the relation of a set and its subsets.

Level: III
Number of players: 1
Materials: None
Procedure: Ask, "Are all of the boys in this class children? Are all of the girls children, too? Are there more girls or more children?"
Note: Do not attempt to teach children to give the correct answer. This activity is intended to keep the teacher informed on the development of each child's logical thought.

One Must Go

Purpose: To help children to notice likenesses and differences.
Level: III

Becky, age 5, identifies the picture that doesn't belong "because she has glasses."

Number of players:	3 to 10
Materials:	Eight mounted magazine pictures:

a dog
a smiling child
a smiling black woman
a smiling white woman
a serious white woman
a smiling white woman wearing glasses
a smiling white man wearing glasses
a man and a woman

Procedure: Allow children to discuss the pictures, and then ask one to "guess' which doesn't belong with the others. Once the dog has been removed, many choices are possible. Allow each child to explain his choice.

Touch and Find

Purpose:	Practice sorting using the sense of touch.
Level:	III
Number of players:	3 to 10
Materials:	Feel Box
	Children's clothing—hat, mitten, sweater, scarf
Procedure:	Have children take turns finding an article in the box that you describe, by saying, for example, "Find something to wear on your head." "Find something you should have two of," and so on.
Variations:	1. Use swatches of fabric, smooth, rough, furry.
	2. Attribute blocks—find two that have the same shape.

Listening Activity

Purpose:	Develop listening skills, reading readiness.
Level:	III
Number of players:	1 to 10
Materials:	None
Procedure:	Allow children to rest their heads on the table and close their eyes. Recite three words, two of which sound alike in some way. Have children take turns identifying the one which doesn't belong: big, dark, boy (initial consonants); train, try, two (initial blends); doll, chairs, kittens (plurals); bent, hot, sleep (final consonants); raining, swing, table (endings).
Variations:	Rhythms—allow children to listen to two lullabies, one marching song.

What Is Not?

Purpose:	To note children's ability to understand negation.
Level:	III
Number of players:	3 to 10
Materials:	Attribute blocks
Procedure:	Display blocks. Give each child a turn to choose a block you describe. As the game progresses allow children to give the directions, which must be in the form: Find a block that is _____ but is not _____.
Variations:	Whenever children are lining up or choosing materials, concepts covered that day can be reenforced: colors, names, sounds, addresses, etc. "If you have a four in your

(a)

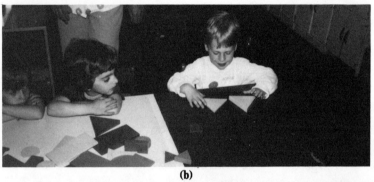

(b)

(c)

Four-year-olds unable to choose a block that is *not* (a) a square, (b) a triangle, (c) a circle.

telephone number but do not have green on today, you may get on line." (Being "first" is a great motivator!)

Letters Alike

Purpose:	Develop reading readiness, recognition of letters of the alphabet.
Level:	III
Number of players:	1
Materials:	An egg carton. Set of letters written on oaktag squares.
Procedure:	Ask child to put those letters that look alike into one hole.
Variation:	Use numeral cards.

Which Go Together

Purpose:	Practice undirected sorting. To notice if a child can form his own sorting plan.
Level:	III
Number of players:	1 at a time
Materials:	Set of animal picture cards including farm animals, pets, circus animals.
Procedure:	Ask, "Put these in piles the way you think they belong."
Variations:	Use pictures of foods ordinarily eaten at breakfast, lunch, and dinner; pictures of children doing seasonal activities; pictures of things that give light, make noise, move.

The Empty Set

Purpose: To see if children perceive the importance of the null set. Most children at this age will put the blank cards to the side and disregard them as they sort.

Level: III

Number of players: 1

Materials: Set of 3" by 5" cards each with a flower, vegetable, or fruit sticker affixed.
Set of 7 or 8 blank 3" by 5" cards.

Procedure: Ask the child to put the cards into two piles the way they belong.

Going to _____

Purpose: Develop ability to notice likenesses.

Level: III

Number of players: 5 to 15

Materials: Attribute blocks

Procedure: Pretend with children that they are going on a trip. Decide together where to go and what to ride on. Explain that each passenger will need a ticket. Pretend that the attribute blocks are the tickets, but only you, the conductor, know which tickets are good for the destination the group has chosen. Show the children three or four "good" tickets, for example, only triangular shaped blocks are valid for this trip. The game begins with a child choosing a block and asking, "May I go to _____?" The conductor answers, "Yes, you may," only if the child has noticed the likeness and selected a triangular block.

To all other choices the conductor replies, "No, you may not go." After each player has had a turn, those who were unsuccessful may exchange their blocks for another try. Other choices in order of difficulty are: large red blocks, thick or circular, small but not rectangular. If a child is chosen as "conductor" ask him to whisper his choice to you.

Tic Tac Toe

Nearing 6, Valerie has no trouble seeing two relationships: *size* and *color*.

Purpose:	To determine which children are able to sort both horizontally and vertically.
Level:	III
Number of players:	2 or 3

Materials: Sheet of paper showing a Tic Tac Toe diagram
Attribute blocks

Procedure: Place on diagram at least five blocks.

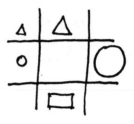

Have children take turns choosing a block to complete the array. Game may be repeated using a different choice and placement of blocks.

Variations: Use set of nine picture cards:
child with blue eyes
child with brown eyes
child with black eyes
men and women with different colored eyes.
Set of any nine picture cards using two properties may be substituted.

Chapter 9

ORDERING

Another important area in the development of mathematical ability is that of order. Nothing is more orderly or sequential than the set of numbers we use for counting. The second grader who concludes that "7 + 8 = 15 because 7 + 7 = 14 and it's one more" is far ahead of his classmates who must memorize each fact independently. An important outgrowth of this discovery is the ability to predict the next term in a series. A third grader correctly "guessed" that 8 x 6 would be 48 because she noticed the following pattern:

$$0 \times 6 = 0$$
$$2 \times 6 = 12$$
$$4 \times 6 = 24$$
$$6 \times 6 = 36$$
$$8 \times 6 =$$

She explained, "The one's number is the number you multiply by and the ten's is half of it." First, she had utilized her ability to classify by disregarding the odd multiples and had then noticed a sequence which proved very helpful.

Younger children often make errors in matters of order since they reach conclusions based on their own sensory experiences. The 3-year-old is big, the baby at home is small. Hence he will say that Daddy is older than grandmother because he's bigger. Children confuse cause and effect. "Today is my birthday because Mother baked a birthday cake." Or, "It is raining because I wore my raincoat."

Another hinderance to understanding order or position is what may seem to a child an inconsistent use of terms. Putting a cup on top of a saucer and putting one's name on top of the paper are two different "tops."

In this section of this book, children will have the opportunity:

1. To understand and properly use terms of order, comparison, and position;
2. To follow directions for ordering according to size, time and number;
3. To order without instruction;
4. To transfer order to a companion set;
5. To discover and continue patterns.

ACTIVITIES

Follow the Leader

Purpose:	To teach words of position: *over, under, through*
Level:	I
Number of players:	3 to 8
Materials:	Set up an obstacle course using large boxes, classroom furniture, and/or playground equipment.

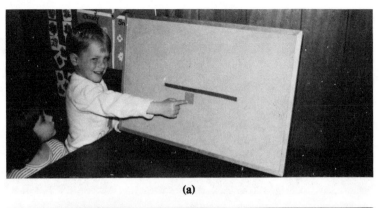

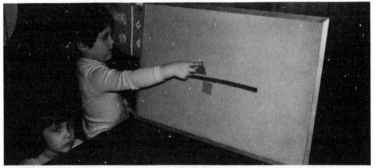

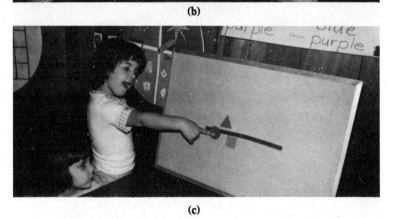

Three and 4-year-olds: (a) *under* (b) *over* (c) *on*.

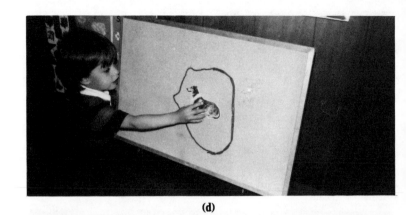

(d)

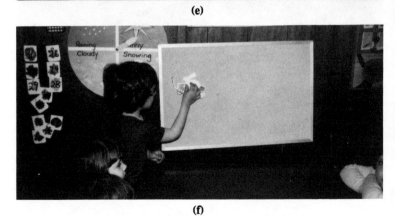

(e)

(f)

Three and 4-year-olds: (d) *inside* **(e)** *outside* **(f)** *in*.

Procedure: Give directions to the leader. As children follow, they call out their position—over the box, under the bar, through the tunnel.

Snack Time

Stephanie chooses the row of crackers that has *more*.

Purpose: To teach understanding of the concepts of *more* and *less*

Level: I

Number of players: 1

Materials: Small snacks such as raisins

Procedure: During snack time set up two rows of raisins:

Allow the child to choose one row. Reinforce words with comments such as, "Sandy, you must really like raisins. You picked the row that has more. This row has less."

Color Cones

Michael, age 3, assembling a color cone.

Purpose: To assist in understanding of order according to number of elements

Level: I

Number of players: 1 to 3

Materials: Three similar color cones

Procedure: Ask children to assemble color cones. Some may need assistance. Next, remove the rings and place one ring on the first, two on the second, and three on the last. Ask, "How did this look when you started." Have one child put the three in order.

Find a Partner

Purpose:	Develop understanding of the concepts of *more* and *less;* setting up a correspondence
Level:	I
Number of players:	Entire group
Materials:	None
Procedure:	Ask the children to form girl-boy partners and line up, boys on one side, girls on the other. Ask questions such as: "Does each girl have a partner?" "Why doesn't Dale have a partner?" "Are there enough boys so that each girl has a partner?" "Are there more boys or more girls?"

Over, Under

Purpose:	Develop listening skills; reinforcing names of body parts; terms: *over, under, below, above*
Level:	I
Number of players:	Group
Materials:	None
Procedure:	Whisper directions to one child, "Put your hands under your chin." Ask next player to "guess" your directions. Game continues as you direct: "Over your head," "Below your ear," "Above your elbow," etc.

Flannel Board Picture

Purpose:	To reinforce the terms: *in, on, over, under, on top, above, below*
Level:	I or II
Number of players:	Small group

Materials: Flannel board set up to represent a bridge across a river.

Felt pieces: cloud, fish, car, boat, airplane, house, flower, etc. (If felt shapes are not available, pictures backed with small pieces of sandpaper adhere nicely.)

Procedure: Have children discuss the scene. Then allow each child to add a flannel piece in an appropriate place: the cloud above the bridge, the fish under the water, the boat on the water but below the bridge, the car on the bridge, the flower on the land, the plane above the cloud.

What Next?

Purpose: Develop understanding of ordering by time sequence

Level: I or II

Tara, age 4, cannot insert those that she missed.

Michael sequencing snowman pictures.

*Number of
players:* 3 to 5

Materials: Picture cards in sets of three: bird building nest, nest with eggs, bird feeding baby bird, plant sprouting, with leaves, with flower (see photo 7a); wrapped gift, child removing wrapping, child lifting toy; double dip ice cream cone, single dip, almost finished.

Procedure: Have children, working together or individually, arrange the pictures in the correct order.

The Three Bears

Purpose: Develop listening skills, ability to order by size; to transfer order to other set

Level: I, II, III

Number of players:	Group
Materials:	Mounted illustrations from The Three Bears Flannel Board
Procedure:	As children listen to the story, have them keep a picture record of the order of events, matching the chairs, bowls, and beds with the appropriate bears. Then have children recount the story using the pictures as a guide.

Guessing Game

Purpose:	Comparing sets: *more, less, same as*
Level:	II
Number of players:	Group
Materials:	A bag of small blocks (11 to 20)
	A bag of beads (more than blocks)
Procedure:	Shake the bags and allow children to guess the contents. Next, ask each player to guess which bag contains more items. Ask, "How will we find out who is right—more blocks or more beads?" Follow the children's suggestions even if they are fruitless. Encourage the suggestion of a one to one pairing. Repeat the conclusion reached by the pairer, "Since we have some beads left over, there are more beads. There weren't enough blocks, so we have less (or fewer) blocks.

Lots of Birthdays

Purpose:	To practice ordering by the number of elements, and transferring order to another set.

Level:	II or III
Number of players:	Small group
Materials:	Pictures of five birthday cakes
	Pictures of children from 1- to 5-years-old
Procedure:	Explain that today is the birthday of each of the children in the pictures, and that each has his or her own cake. Have a child draw one candle on a cake, the next player draws one more candle on the next cake, and so on. Mix up the pictures and have them rearranged on the "one more than" principle. Finally, have the players match each child with his or her cake.

I Spy

Purpose:	To reinforce terms of position
Level:	II or III
Number of players:	Small group
Materials:	None
Procedure:	Choose a child to begin the game. Have her whisper to you something she sees in the room. After each hint describing its position, the other players may guess. The first to name the object becomes "it." Encourage use of: *above, below; over, under; on top of, underneath.*

Musical Chairs

Purpose:	To reinforce concepts of *not enough, too many, less*
Level:	II or III

Number of
players: Group

Materials: Chair for each child, arranged in reversed positions.

Music—recording or piano.

Procedure: For the first round, have a chair for each child. Remove a chair, stressing that you now have one less chair, too many children, not enough chairs. Repeat for each round until there is a winner.

Left (or Right) Day

Purpose: To introduce the concepts of *left* and *right*

Level: III

Number of
players: Entire class

Materials: Decals, ribbon bracelets, or pipe-cleaner rings

Procedure: After introducing the terms *left* and *right*, choose Left to stress today. Place the decal on each child's left hand. (These will wash off but cannot be removed or transferred by the children.) Reinforce throughout the day: raise your left hand, hop on your left foot, touch your left ear, etc. A few days later, have a Right Day."

Looby Loo

Purpose: To reinforce concepts of *right* and *left*

Level: III

Number of
players: Entire group

Materials: None

Procedure: Have children arrange themselves in a row, and while singing "Looby Loo," act out the directions:
"I put my right foot in,
I put my right foot out,
I give my right foot a
shake, shake, shake
And turn myself about."
Repeat for left foot, other parts of body.

Huckle Buckle Beanstalk

Purpose: To reinforce concepts of *right, left; higher, lower*

Level: III

Number of players: Entire class

Materials: None

Procedure: Choose one child to hide her eyes, and another to select an object in the room. The person who is "It" questions each child in turn and they must answer "higher," "lower," "left," or "right" until the object is guessed.

Grid Game

Purpose: To develop the ability to follow directions involving position: *left, right; top, middle, bottom*

Level: III

Number of players: 3 to 8

Materials: A grid containing nine squares, three across and three down, for each child.
Nine different stickers for each player.

Procedure: Have children place stickers according to your teacher's directions: "Find your blue star and put it in the bottom, left corner." Point to the top, left corner. "Paste your red heart there." Continue until each position is filled.

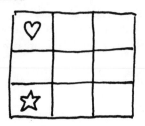

White, Red, or Green?

Purpose: To familiarize children with Cuisenaire Rods; to practice ordering according to length

Level: III

Number of players: 4 to 6

Materials: Cuisenaire rods

Procedure: Allow children to play with the rods for a few minutes, and then ask them to each choose a white, a red, and a light-green rod. Next, have each child guess which of the three rods you have placed in his hand, which is held behind his back. In order to be successful, the felt length must be compared to the length that is seen. Game continues with two children as partners.

Build a Staircase

Purpose: To help children learn to order by length

Level: III

Number of players:	4 to 6
Materials:	Cuisenaire Rods
Procedure:	Ask children to choose one rod of each color and to arrange them "the way they belong." Show those children unable to order without direction, an assembled staircase:

Michael building a Cuisenaire staircase.

saying, "fix yours like this." Some may need more specific instructions. "find the longest rod. Find the next longest and place it over the other," and so on.

Which Fits

Purpose:	To develop the concept of transferring order
Level:	III
Number of players:	1 to 3
Materials:	Dolls from doll corner Dolls' clothing

Procedure: Have children order the dolls from smallest
to largest and dress smallest doll with smallest clothing, etc.

Flannel Board Story

Purpose: To give children an opportunity to practice
transferring order; classification

Level: III

*Number of
players:*   3 to 4

Materials: Flannel board
Set of flannel pieces of dogs, doghouses,
bones, puppies, (commercial set available
from *Creative Playthings*)

Procedure: After a general discussion of the pictured
dogs, ask one child to order them according
to size. Show an equal number of houses and
allow another child to match them appropri-
ately. Follow with bones and puppies.

Who Sits Where?

Purpose: To practice ordering according to the number of elements

Level: III

*Number of
players:*   3 to 5

Materials: Pictures (from a catalog) of tables with from
0 to 8 chairs

Procedure: Say to children, "Pretend that your mother
is going to get a new set of table and chairs.
Which do you think would be good for your
family? Remember that each person needs a
chair." Allow children to make a choice and
to tell who would sit in each chair.

Noticing likenessess—matching puppies with "mommies."

Now say, "There is one picture that doesn't have any chairs. Let's put that first. Which do you think we should put next?" Continue ordering in like manner. Counting may be reinforced by naming the number of chairs in order.

Getting Dressed

Purpose:	Practice ordering
Level:	III
Number of players:	1 to 4 or 5
Materials:	Colorforms: board, dolls, clothing
Procedure:	Display doll on board. Allow children to choose the articles of clothing that would be appropriate to the day's weather. Have children order the clothing in the dressing sequence. Many orders would be appropriate, but socks must come before shoes, which come before boots, etc. After clothing has been arranged in a line, have children dress the doll to check the order.

Seasons

Purpose:	To introduce repeating order
Level:	III
Number of players:	Small group
Materials:	Pictures of seasonal activities which children have found in old magazines. Bulletin board
Procedure:	Discuss with children the four seasons of the year. Have them sort the pictures into sets

for each season. Display on the bulletin board in clockwise order, with the current season on the top. Reinforce through discussion, stressing that season follows season year after year.

Adaptation: Depict special events occurring in the classroom and over the weekend using the days of the week.

Stringing Beads

Purpose: To teach children to duplicate a pattern
Level: II
Number of players: 1 to 3
Materials: Wooden beads, yarn or shoe laces for stringing

Valerie continues a block pattern by color.

Procedure:	Ask each child to duplicate the pattern you have strung. Possibilities, arranged in order of difficulty: red, yellow, red, yellow . . . ; cube, cube, sphere, cube, cube, sphere . . . ; small, larger, largest, small, larger, largest . . . ;
	green, blue, red, red, green, blue, red, red . . .; yellow, green, yellow, blue, yellow, red . . . ; cube, sphere, cube, sphere, sphere, cube; sphere, sphere, sphere. . ..
Variations:	Use pegs on a pegboard to form patterns. Use star stickers on paper.

Rhythms

Purpose:	To develop listening skills: ability to hear a pattern
Level:	II
Number of players:	Small group
Materials:	None
Procedure:	Have children listen as you clap a rhythm. When they think they can duplicate your rhythm, ask them to join in.

Move About

Purpose:	To develop the ability to feel a pattern using large-muscle coordination
Level:	II
Number of players:	Small group
Materials:	None
Procedure:	Have children repeat and continue patterns, such as two jumps, one clap, twirl about;

one step in, one jump, one step out, clap above the head.

Variation: Using electrician's tape for durability, tape a number line on the floor. Do not label the steps with numerals. Children follow the leader's pattern. Example: one step, one jump, two steps, one jump, three steps, etc.

Follow the Bus

Purpose:	Continue practicing patterns on a grid
Level:	III
Number of players:	Small group
Materials:	Small toy bus, grid pattern drawn on the board or floor
Procedure:	Have children pretend that the grid is a picture of streets and that the bus travels along a pattern—two blocks east and makes a stop, one block north and makes a stop, two blocks east, etc. Choose one child at a time to move the bus to its next stop.
Variation:	A bee on the ceiling.

Reinforcing Activities

Level:	III
Number of players:	Small group
Materials:	Dittoed work sheets
Procedure:	Have children mark "the one which comes next"

Example:

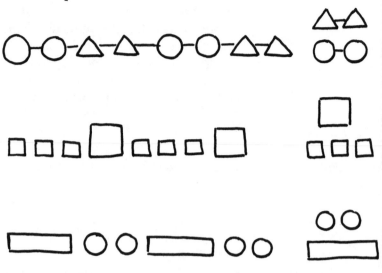

Variations: Those children who have no difficulty with the examples above may enjoy drawing figures to continue a pattern:

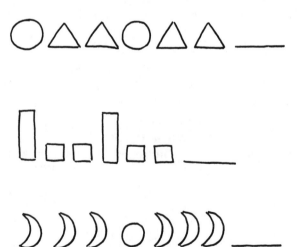

Chapter 10

COUNTING

WHAT IS COUNTING?

A second grade teacher was shocked when some of her students were unable to count out 12 sheets of paper for booklets they were making. She thought she had taught them to "count" to 1000. Repeating the names of the numbers in proper order is called *rote counting* and, while an important first step, it is not very functional. The ability to answer the question, "How many?" is called *rational counting,* a complex process too often taken for granted by parents and teachers.

Many skills are needed before a child can really learn rational counting. First, the child must be able to set up a correspondence between each number name and one of the objects to be counted. In the beginning this is a physical action. The child touches each element of the set as he recites the number words. During kindergarten most children are able to count with some distance between themselves and the set but continue to point at each element. Nods of the head seem to suffice in the upper grades. At age 12 most children can perceive the number of a set with eight or fewer elements at a glance.

After learning this process of correspondence, the child must know when to stop. When she has given each object one number name, she must not continue saying the numbers nor touch any object again. When counting the children in a group it is not unusual for a child to "count" herself twice. Counting figures which have been placed in a circular order or drawn randomly on paper is troublesome. Since there is no apparent beginning or end, children are unable to devise a plan and tend to bounce around missing some while touching others many times.

Finally the child must realize that the last number named counts the set. Teachers should remember to ask "How many are there?" and have the child answer, "Five."

Another abstract concept involves *transitivity,* the realization that two sets that are in a one-to-one correspondence have the same number. One cracker is placed in front of each child and the teacher asks, "How many children are here?" After counting, a child responds, "Six." To be able to transfer the six to the number of crackers requires a degree of logical thought which many 5-year-olds have not attained. Typically, children will recount to determine the number of crackers.

Uses of Number

Because numbers are used in a variety of ways, children may become confused. Teachers should be aware of these uses and avoid situations which might be misleading:

Cardinal Numbers. Numbers used in the cardinal sense have been discussed. Cardinal numbers count a set. They answer the question, "How many?"

Ordinal Numbers. As the name implies, ordinals indicate order. Often these are said as first, second, third,

. . . However, when a child is told he is five on line, it does not mean that he is five children. Five is used here in the ordinal sense, meaning he is fifth.

Labels. Channel nine is not nine channels nor is it even the ninth channel. Nine is simply a designation or name.

Position. Room 213 probably means the thirteenth room on the second floor.

Through experience the child learns to accept these variances. In the early learning stages it is best to use number in the cardinal sense only.

Teaching Children to Count

Counting instruction, like anything else, begins where the child's ability ends. The alert teacher will determine each child's level and begin instruction at the appropriate place. A few minutes of individual attention is far more valuable than time spent having the class parrot the counting numbers. Those children who know the material dominate the recitation while very little is accomplished by those who do not.

The way a child responds when asked to get one (book, carton of milk, box of crayons) for each child at his table is an indicator of the level of his perception of counting. Robbie brings "alot" and allows each to fend for himself. He becomes puzzled if he has brought too few or too many. This shows a lack of understanding of a one-to-one matching. Meg makes many trips, bringing one or two at a time. She understands the one-to-one relationship but is unable to use counting as a functional tool. Lee uses the children's names instead of the counting numbers, "One for me, one for Craig, one for Beth." This may indicate that she does not know the names of the numbers in order or that she is not able to transfer the number of children to the number of objects. Of course, children who are able to count functionally, count the children and return with the proper number of objects.

Most 3-year-olds understand the concept of "one." They have heard the word used and its meaning demonstrated each time Mother has said, "only one." They may not, however, think of it as the first of a series of numbers. "Three" is also familiar, usually well established by this age. Many children recognize a set of three elements without counting. Psychologists explain that there are three important people in the child's world: himself, Mommy and Daddy. He has been taught to answer "three" and to put up three fingers when asked his age. He was too young to do this last year. Most of the stories and poems he hears involve three characters: Three Bears, Three Little Kittens, Three Blind Mice, Three Little Pigs, and on and on. Finding the lost boot or mitten is an opportunity to illustrate one to one correspondence while stressing "two." Each of the succeeding numbers are introduced, one at a time, gradually increasing the child's ability to count rationally to ten.

To test rote counting, four kindergarteners were asked to "count for me." Their responses:

Gary: "1, 2, 3, 4, 5, 6, and that's all."

Anita: "1, 2, 3, 4, 5, . . . 23."

When asked what came next, she shrugged her shoulders.

Mark: "1, 2, 3, 4, . . . 11, 12, 50, eleventy."

Scott: "1, 2, 3, 4, 5, 6, 7, 8, 9, 10, 20, 30, 100."

None of these are serious problems. Gary's instruction will progress to seven. The other answers indicate that the children do not understand the orderliness of our number system. Only after mastery of rational counting through ten should place value and the hundred chart be introduced.

The ability to count rationally is improved when there is a need to know "how many?" The following activities include suggestions for counting lessons that are meaningful to children.

One for Each

Purpose:	To develop the concept of a one-to-one correspondence
Level:	I
Number of players:	1
Materials:	No specific materials. Take advantage of every opportunity to provide practice for the children.

"One for each."

"One for each."

Procedure: Say to a child, "Please give one straw (or container of milk, or cracker) to each child." or, "Put one paint brush in each jar of paint."

"Plant one seed in each container."

"Put one toy in each cubby."

"Drive one car into each parking space."

Fill 'Em Up

Purpose:	Develop understanding of one-to-one correspondence
Level:	I
Number of players:	2
Materials:	A large egg carton (24-egg capacity) with a line drawn down the center
	10 or 12 envelopes containing anywhere from 0 to 4 cardboard discs

Procedure: Have each child in turn choose an envelope and place each "egg" into his side of the carton. The first to have one egg in each hole is the winner.

Variations: To increase the difficulty, a "start" and a "finish" point may be designated with two paths drawn on the carton.

How Many Is Two?

Purpose: To reinforce the meaning of *two*

Level: I

Number of players: 3 to 5

Materials: None

Procedure: Teach the song "I Have Two Eyes to See With" and have children act it out.

Variations: Many other songs and finger plays provide interesting repetition of the concept of *two*.

One, Two, Three

Purpose: Develop concept of rational counting to three; reinforcing colors and shapes

Level: I

Number of players: 2 to 4

Materials: Attribute blocks

Procedure: Allow each child to choose three blocks. Have them recite the number names with you if necessary. Ask each player how many blocks she has. Ask appropriate questions of each child. "How many are blue? How many are circle shapes? How many are large? Who has the most red blocks? Who has two square shapes?" etc.

Which Is More?

Purpose:	To reinforce the concept that two is less than three
Level:	I
Number of players:	Group
Materials:	Animal crackers or other small snacks
Procedure:	Tell children that in this napkin you have wrapped two crackers and in the other, three. Allow each to choose which he wants.

Guess How Many

Purpose:	Develop concept of rational counting to five
Level:	II
Number of players:	2 to 6
Materials:	Feel box or cloth Small objects
Procedure:	Choose a child and have her place 2, 3, 4, or 5 similar objects into the feel box (or under the cloth). Then ask her to choose another player to identify by feel the type and number of objects.

Dominoes

Purpose:	Develop concept of rational counting to six
Level:	II
Number of players:	4 to 6
Materials:	Large cardboard dominoes constructed by placing the dots in different positions
Procedure:	Seat children in a circle. Have each player choose six dominoes from the pack, which is

placed face down on the floor. One child begins the game by placing one domino in the center of the circle. The player to her left must match either side of the domino. If he does not have a match, he must draw from the pack until one is found. The winner is the first to use all his dominoes.

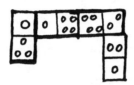

Around We Go

Purpose: To reinforce counting
Level: II
Number of players: Entire group
Materials: None
Procedure: Have children form a ring and march around to music. Call out, "Beginning with Cindy, seven." Cindy begins the counting. The player who says "seven" goes inside the ring and the others resume marching. The last player to remain on the ring is the winner.

Scavenger Hunt

Purpose: Practice rational counting
Level: II or III
Number of players: 4 or 5
Materials: Prepare a chart picturing varying numbers of objects that can be found in the room (or play yard): blocks, books, crayons, etc.

Procedure:	Have children collect the correct number of each item pictured. The winner is the first to finish.

Counting Animals

Purpose:	Practice rational counting
Level:	II or III
Number of players:	Entire group
Materials:	Card for each player with from one to ten dots on each
Procedure:	Have all the children pretend they are animals. As you call out a number, the child holding the card with that number of dots on it steps forward and imitates the animal he is pretending to be the appropriate number of times, using sounds or actions. The other players try to identify the animal.

Roll and Jump

Purpose:	Practice rational counting
Level:	II or III
Number of players:	2 or 3
Materials:	One die
Procedure:	Designate a starting point and finishing line. Have each player in turn roll the die and take the appropriate number of jumps. The first to cross the finish line is the winner.

Board Games

Purpose:	Practice rational counting
Level:	II or III
Number of players:	2 to 4

Materials: Any simple commercial board game using one die

Procedure: Simplify the rules of each game to match the ability of the players.

Peanut Hunt

Purpose: Practice rational counting
Level: III
Number of
players: Entire group
Materials: A bag or two of peanuts
Procedure: Have each child count the peanuts she has found and then compare numbers to determine the winner.

Hide and Seek

Purpose: Practice rational counting
Level: III
Number of
players: Entire group
Materials: None
Procedure: Divide class into two groups. Have one group hide, and the second group find them, but the seekers must count the hiders before and after to make certain that each has been found.

Spinner

Purpose: Practice rational counting
Level: III
Number of
players: Entire group
Materials: Divide a game board into fourths, identify each with a cartoon character and place a spinner in the center.

Procedure: Have children form four teams and ask each team to choose one of the cartoon characters. Each child spins and records the result with a tally mark. Tallies are counted to determine the winning team.

Our Pets

Purpose: Practice rational counting
Level: III
Number of players: Entire group
Materials: Bulletin board prepared with a grid for a simple bar graph
Stickers of an appropriate size (mailing labels are fine)
Procedure: Following a discussion of the children's pets, have each child take one sticker for each of his pets and paste it in the proper bar. Each column is then counted. The graph affords many opportunities for comparing numbers.

Variations: Similar graphs may be constructed using other data: favorite TV show, number of children in the family, eye color, number of missing teeth.

Birthday Cakes

Purpose: Develop concept of ordinal numbers
Level: I
Number of players: 4
Materials: Picture cards of birthday cakes with 1, 2, 3, and 4 candles
Picture cards of children whose ages correspond (dolls may be used)
Procedure: After discussing the cakes and counting the candles, have children arrange them in order. Stress that "On your first birthday you are 1-year-old" and ask a player to choose the doll which corresponds. Proceed with emphasis on the relationship of second and 2.
Note: While every child seems to know the meaning of "first," the words "first" and "second" give no oral clue to their relationship to "one" and "two". The above activity stresses this correspondence.

Listen and Do

Purpose: Develop concept of ordinals; improve listening skills
Level: II
Number of Players: Group
Materials: None
Procedure: Begin the game by explaining that you will give three directions while the players listen.

When you say "Go," the children must follow the directions in the order given. "First, touch your head. Second, clap your hands. Third, touch the floor." Children may take turns being the leader. As listening skills improve, more directions may be included.

Days of the Week

Purpose:	Develop concept of and teach ordinals; the days of the week
Level:	II
Number of players:	Group
Materials:	Long strip of activity paper divided into seven sections, each representing a day.
Procedure:	Have children discuss Sunday activities and contribute pictures (original or from magazines). This may be an on going project completed over a period of weeks. The stress throughout is that Sunday is the first day of the week; Monday, the second; up to Saturday, which is the last.

Finger Plays

Purpose:	Practice ordinals
Level:	II or III
Materials:	None
Procedure:	Many familiar finger plays provide interesting practice.

Following Directions

Purpose:	Practice ordinals, listening skills
Level:	II or III

Number of	
players:	Any number
Materials:	A "ditto," showing a mother duck followed by five ducklings, using a left to right progression, for each child.
Procedure:	Ask children to follow your directions. Then say, "Color the third duck yellow. Put an x on the first baby duck. Draw a ring around the last duck," etc.
Variation:	Have "ditto" show children on line and give appropriate directions: "Color the second boy's shirt red, color the fifth child's eyes blue," etc.

Going Shopping

Purpose:	Practice ordinals
Level:	III
Number of	
players:	5 to 10
Materials:	Poster or flannel-board display of a department store with 5 or 6 floors
	Many small pictures of various items
Procedure:	Decide together which floor will sell toys, clothing, tools, etc. Have children take turns affixing various items, while repeating the ordinal. For example, say, "A chair goes with the furniture on the fourth floor."
Variation:	Depict an apartment building with a movable elevator. Have children put a paper doll on the elevator and say, "This is Devon. He gets off on the third floor."

Chapter II

NUMERALS

Numerals are the symbols which are used to represent numbers. The abstract configuration "5" brings to mind a set of five elements: fingers, apples, or whatever, only because the observer has learned to read the symbol. It stands to reason that children are ready to learn numerals at the same time that they can learn the letters of the alphabet. Numerals should be introduced only after a thorough understanding of the cardinality of sets is achieved.

The activities in this chapter follow a developmental sequence which has been designed to minimize problems that children often have when writing the numerals: reversals, writing the wrong numeral, producing unrecognizable scribbles. A first step is acquiring the ability to recognize symbols which are alike. A certain degree of visual perception is necessary to notice that 7 is not the same as Γ . If a child perceives these as exactly alike, she cannot be expected to write consistently correct sevens. Next, she learns to read the symbol, to associate a familiar word to the numeral. Finally, after many opportunities to copy and trace the figures, she is ready to write. Teachers should be reasonably sure of success before asking a child to write a numeral. The amount of readiness varies with the individual. Time spent preparing a child for success is more enjoyable than its alternative, remediation. Size, spacing, or staying within guidelines are not important at the introductory level. While a certain amount of drill is necessary, filling a page with twos can be a painful chore or, at the least, a boring experience.

Each numeral should be introduced after the preceding one has been mastered. Teachers should be consistent in their formation of the numerals and must be alert that children are making the proper strokes. Manuscript numerals are used until children learn cursive writing, usually in second grade.

script   1 2 3 4 5 6 7 8 9
cursive   1 2 3 4 5 6 7 8 9

Some children will be able to progress through these steps very rapidly. Others may not be ready to write the numerals until first grade.

ACTIVITIES

Find Your Partner

Purpose:	Practice matching like numerals
Level:	III
Number of players:	Any even number up to 18
Materials:	Two sets of numeral cards
Procedure:	Distribute the numeral cards (making certain there are two of each) face down. At the signal, children move about finding a match.

Match the Numeral

Purpose: Practice matching like numerals
Level: III
Number of
players: 5 to 10
Materials: A set of numeral cards for each player
Procedure: Have the children watch as you write a numeral on the chalkboard. The child holding the matching card holds it up.

Numeral-Go-Round

Purpose: Practice matching like numerals
Level: III
Number of
players: 1
Materials: A circular-shaped cardboard with the numerals 1 through 9. Clip clothespins with the numerals printed on them.
Procedure: Player clips the clothespin to the chart matching the numerals.

Which Are Alike?

Purpose:	Develop directional visual discrimination
Level:	III
Number of players:	1 (A ditto may be used for a small group.)
Materials:	A card with the numeral printed on the top correctly; the same numeral printed below correctly and incorrectly

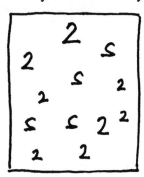

Procedure:	Ask the child to point to each figure that looks just like the one at the top (size is not important).

Sorting Box

Purpose:	Practice matching like numerals
Level:	III
Number of players:	1
Materials:	Sorting box labeled with numerals Pack of numeral cards
Procedure:	Child sorts cards by matching numerals. Cards for which there are no matches may be placed in a separate pile.

Find Your Name

Purpose: Practice reading numerals
Level: III
Number of players: Group
Materials: Numeral cards displayed around the room
Procedure: Give each of the children a numeral name, then ask them to find the numeral and stand by it. Numeral cards may be hidden to add interest.

Clothespin Soldiers

Purpose: Learn sequencing of the numerals
Level: III
Number of players: 1
Materials: "Soldiers" made by sketching a face and a numeral on a wooden clothespin.

Procedure: Have child arrange "soldiers" in order on a line.

Feel a Numeral

Purpose: Develop numeral recognition
Level: III

Number of players:	3 to 10
Materials:	None
Procedure:	"Write" a numeral on a player's back with your finger. Ask child to name the numeral you wrote.

Twister

Purpose:	Practice sequencing of the numerals
Level:	III
Number of players:	3 to 5
Materials:	Large numeral cards
Procedure:	Place the numeral cards on the floor at random. Have players take turns stepping on them in order.

Go Fish

Purpose:	Enhance numeral recognition
Level:	III
Number of players:	3 or 4
Materials:	Deck of playing cards. Face cards may be removed.
Procedure:	Give each player seven cards, which are held or placed out of view of the other players. Place the rest of the deck face down in the center. A player begins by asking the person on his left for a card which will match one in his hand. If the second player has the card he must give it up. If he does not he says, "Go Fish." The first player draws from the deck until he finds a match. When all four cards of any number are col-

lected they are placed on the table. The first person who has no cards left in his hand is the winner.

Variation: The card game "War" reinforces the *less than* and *greater than* relationships.

Who Am I?

Purpose: Develop numeral recognition; *less than, more than*

Level: III

Number of players: 4 to 8

Materials: Numeral cards with strings attached

Procedure: Place a numeral card on the player's back. Ask her to try to guess which numeral she is wearing by questioning the other players who must answer only *more, less, or yes.*

Dot to Dot

Purpose: Enhance recognition of numeral order

Level III

Number of players: 1

Materials: Follow-the-dot sheets. These may be made by placing a sheet of slightly transparent paper over a simple picture.

Procedure: Have child connect the dots in proper order to form a picture.

Numeral Puzzles

Purpose: Help children learn to associate the numeral to a set

Level: III

Number of players: 1

Materials: Set of puzzles (these are available commercially or may be teacher made)

Procedure: Have child match the numeral to the set. These puzzles are self correcting.

Dominoes

Purpose: To help children associate the numerals with sets

Level: III

Number of players: Small group

Materials: Dominoes with set of dots on one half, a numeral on the other

Procedure: Have children take turns matching a dot side to a numeral side.

Bulletin Display

Purpose: Teach children to associate numerals with sets

Level: III

Number of players: 1

Materials: Chart with numerals (ribbons attached) and sets in random order.

Procedure: Have child match each numeral to a set.

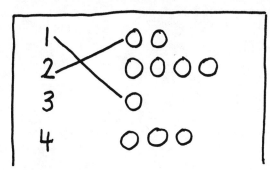

Sticker Fun

    Purpose: Teach children to associate numerals with sets

    Level: III

    Number of players: 3 to 6

    Materials: Numeral sheet for each player
                Stickers
                Spinner

    Procedure: Have player begin by spinning the spinner and reading the numeral to which it points. He counts out the correct number of stickers and pastes them next to the numeral on his card. The first player to complete his card is the winner.

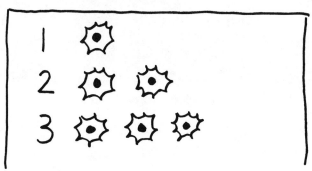

Matching numerals to sets.

Number Train

Purpose:	Develop association of numerals with sets
Level:	III
Number of players:	This may be a class project or may be done individually.
Materials:	Pictures of an engine and boxcars Small pictures to fill each car
Procedure:	Have children number each boxcar in order and place the appropriate number of items in it. As each new numeral is learned, another car is added.

Find a Pair

Purpose:	Develop association between numerals and sets
Level:	III
Number of players:	2 to 4
Materials:	Make a playing deck using twenty-one 3" x 5" cards. Ten contain the numerals O through 9. Ten show sets of 0 to 9 elements. One is the lucky card.
Procedure:	Deal all cards. Have one child draw a card from the player on her left and put down all of her pairs. Continue in like manner. The player holding the lucky card after all of the pairs are found is the winner.

Concentration

Purpose:	Develop association of numerals with sets
Level:	III

Number of
players: 2 to 4
Materials: Cards containing the numerals 0 through 9
Cards showing sets of 0 through 9 elements
Procedure: Place cards face down in two rows, in random order. Have first player turn two cards up, attempting to find a pair. All players should look at the cards, and try to remember the positions, before they are placed face down again. When a player finds a pair of matching cards, he may choose two more cards. The winner is the player with most cards.

Getting the Feel of the Numerals

Purpose: Develop numeral writing readiness
Level: III
Number of
players: 1
Materials: Tactile numerals cut from fake fur, mounted on tagboard
Procedure: Have child trace the numeral using the proper strokes.

Tracing Numerals

Purpose: Develop numeral writing readiness
Level: III
Number of
players: 1
Materials: Numeral cards
Tracing paper
Procedure: Allow child to trace the numerals.
Variation: To add interest, allow child to trace over carbon paper.

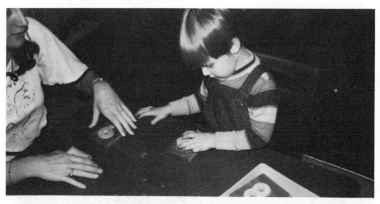

Getting the feeling of the numerals.

Coloring Numerals

Purpose:	Develop numeral writing readiness; practice reading numerals and following directions
Level:	III
Number of players:	Small group
Materials:	Numeral sheets, crayons
Procedure:	Tell children to follow your directions, then say, "Color the *five* green."

Stencils

Purpose:	Develop numeral writing readiness
Level:	III
Number of players:	1
Materials:	Numeral stencils
Procedure:	Have child trace through the stencil.

Find the Numerals

Purpose:	Practice numeral writing readiness
Level:	III
Number of players:	Group

Valerie finds the hidden numerals.

Materials: Duplicate simple pictures containing "hidden" numerals.

Procedure: Have children trace over the numerals.

Practice Before Paper

Purpose:	Practice numeral writing—unhappy results disappear quickly!
Level:	III
Number of players:	1
Materials:	Container (frozen vegetable box is a convenient size) filled with sand or salt, or

"Magic Slate," or
Chalkboard

Procedure: Allow child to "practice" writing numerals in sand or salt, or on "Magic Slate" or chalkboard before attempting more permanent results.

My Number Book

Purpose: To culminate counting, numeral activity
Level: III
Number of players: Individual project
Materials: Construction paper
Loose-leaf binder rings
Procedure: Have each of the children prepare a booklet containing numerals they have written associated with appropriate sets.

Chapter 12

READINESS FOR ADDITION

Addition is based on counting. When a set of four elements is joined with a set of two elements the new set contains six elements. From this union it is concluded that $4 + 2 = 6$. In the past, educators assumed that once children had learned to count they were ready to begin addition. Piaget has found, however, that a certain degree of logical thought is necessary for the operation of addition to be meaningful. Young children have been memorizing addition facts for years, but, to them, these facts are completely unrelated. Many children must wonder why it took so long to learn $3 + 4 = 7$ in first grade when by second grade they are able to reason that $3 + 4$ must be one more than $3 + 3$. Those who were compelled to memorize mathematics before their logical development allowed them to see relationships grew up thinking that all mathematics was rote memorization.

Piaget maintains that the most persuasive indicator that children are ready to learn mathematics on other than a rote basis is their ability to conserve. His research shows that approximately 70% of 7-year-olds are conservers, ready to begin the study of mathematics on the concrete or manipulative level. When children reach this concrete operational stage, their reasoning enables them to classify and to order, both necessary concepts for meaningful mathematics learning. Children who do not yet conserve may be given ample opportunity to join sets and count the elements in the resulting set, but no pressure for memorization should be applied.

One teaching aid, the Cuisenaire rods, seems to circumvent the restrictions of Piaget. These are a set of wooden sticks, ranging in length from one centimeter to ten centimeters, that could be described as "numbers in color." Rods of the same length are the same color. Children 4 and 5 years of age are able to sort these by color and discover that the rods in each "pile" have the same length. They can then refer to "the green rod" with assurance that it is the same length as every other green rod. Building a "staircase" or ordering the lengths can be accomplished with little or no direction.

While playing with the rods, many children soon discover that two rods, laid end to end, have the same length as one other rod.

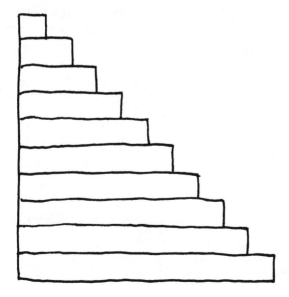

With encouragement by the teacher, the child finds two others that "fit" along the yellow rod.

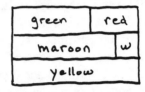

Since the length remains constant, even the nonconserver comprehends that red and green is the same as white and maroon. Later, when the rods are given number names, this is translated to $2 + 3 = 1 + 4 = 5$. In attempting to find another two-rod "train" to fit under the yellow rod, the child places a maroon rod followed by a white and concludes that this is the same as white and maroon, the commutative property of union. By allowing three- and four-rod trains she may notice that $1 + 1 + 3 = 2 + 3$ or $1 + 4$, the associative property of addition.

Once again, the set of rods is a manipulative learning aid. Your role is to suggest through questioning. Regardless of how apparent a conclusion or generalization may be to you, do not attempt to force this upon the learner.

ACTIVITIES

Fish

Purpose:	Practice union of sets; addition readiness
Level:	III
Number of players:	4 or 5

Materials: Bag of fish crackers
One piece of blue paper for each player

Procedure: Have each player place his or her "ocean" on the table. Place from one to five fish on each paper, and explain that today will be "five fish day." Ask the players to tell how many more fish they need to make five. Next, direct the players to move the fish about while you ask questions, for example: "Two fish swim to the top of the ocean while the rest swim to the bottom. How many are on the bottom? One more swims to the top. Now how many are on the top? All except one fish dives to the bottom and along comes a great big fish (the child) and gobbles up the one on the top. How many fish are left?" The activity continues until all the fish are eaten.

Variation: Animal crackers in a circus cage may be used.

Make a Set

Purpose: Practice numeral recognition, counting, joining sets

Level: III

Number of players: 2

Materials: Set of uniformly sized blocks—one block with the numerals 1 to 6 on the faces

Procedure: Have first player toss the block, read the numeral, count out the appropriate number of blocks and place them in a line. The player then throws again, placing the second set of blocks with the first and counts to find the total. Have the second player do the same thing and place his blocks under the first set. The player with the most blocks keeps all of the blocks, the player with the most blocks at the end of the game is the winner.

Matho

Purpose:	Practice addition readiness
Level:	III
Number of players:	4 or 5
Materials:	Player cards on which the numerals being practiced have been written.

	Caller cards, folded in half with appropriate sets of dots on the outside.
Procedure:	Draw a caller card and show one side to the players. Have children count the dots, and then show them the other side of the card. Ask children to cover the numeral which names the sum. The first player to cover his card is the winner.

Computer

Purpose:	Practice reading numerals; readiness for addition
Level:	III
Number of players:	2

Materials: Large box for a child to sit in with an "input" slit and an "output" window cut into the front.
"Input" cards with sets of dots
Numeral cards for "output"

Procedure: Give the child who is the computer the numeral cards. Have the other player insert two input cards. After making appropriate noises, the computer shows the numeral card which names the sum.

Going Fishing

Purpose: Practice addition readiness
Level: III
Number of players: 3 or 4
Materials: A large box lid for the fishpond
A fishing pole with a magnet for a hook
Many cardboard fish with paper clip mouths on which two sets of different colored spots have been drawn.

Procedure: Show children how to "catch" a fish with the magnet. Ask them to count the spots on the fish they "catch" and then count the total number of spots. Children who count correctly may keep the fish. When all of the fish have been caught, the player with the most is the winner.

Chapter 13

GEOMETRY

The earliest physical experiences of children are geometric in nature rather than arithmetic. Long before children become aware that there are "many" beads on the playpen, they notice that they can push them together, spread them apart. They learn that a ball will roll before they can count three balls. Traditionally, the first geometric instruction that children received was the identification of plane figures. Popular television programs expose children to the vocabulary of geometry before they are able to pronounce the words correctly. "Wectangoo" is a common preschool word! Circles, triangles, and squares are examples of rigid geometric shapes, but Piaget has found that children understand the basic ideas of topology, a branch of geometry not concerned with rigidity, before they understand these Euclidean concepts. At 4 or 5 while unable to draw a figure approximating a circle, most children are able to draw a line that encloses a figure. Some of the activities in this chapter will help to develop other basic topological ideas: *inside, outside; open, closed;* and *between.*

Measurement is another branch of geometry commonly introduced during the preschool years. All measurements are comparisons. Two children stand back to back in direct comparison and the conclusion "that Sharon is taller" is readily apparent. Later in the grades, children will be taught to recognize and use standard units for comparison. For example, they will be able to conclude that a line segment is longer than a decimeter; in fact, it is just about three decimeters long. While this process seems simple enough to adults, it requires a degree of logical thought that has not developed in the preschooler.

Most 4- to 6-year-olds respond correctly that two strips of paper are the same length when they are placed side by side; however, when one is slid upward, children conclude that it grew longer. It appears to them that the strip shrinks and stretches as it is moved about. Piaget found that when water is poured from a short wide container into a tall narrow one, children conclude that there is now more or less water because "it is fatter" or "it is higher." Obviously, formal measurement

Alex, age 4, is certain that the strip on his right is now longer.

for these children is meaningless. If these simple exercises are used as tests of readiness, it is found that most children should begin the formal study of measurement in second grade, or even later.

However, the preschool mathematics program should include readiness for measurement. In addition to the vocabulary of position discussed previously, terms of comparison are introduced: *heavier, lighter; earlier, later; longer, shorter; warmer, colder.* As children learn best through phsycial experiences on the concrete level, provide objects which can be lifted, so that children can experience weight. Allow children to feel two pie tins at room temperature and compare them after one has been left in a sunny place. Instead of attempting to teach children to tell time, concentrate on the concept of time: *morning, afternoon, night; yesterday, today, tomorrow.*

Another readiness topic is the identification of measuring devices by use. In place of the usual yardstick and measuring cup, obtain metric materials if possible, as these will be the units of adulthood. Allow children to handle and experiment with these instruments. Plastic liter containers may be left at the sand table. Children can step on scales and experience the turning and stopping of the dial. A meat or candy thermometer can be observed in warm water. A child can watch the movement of a second hand.

ACTIVITIES

Guess Who?

Purpose:	To reinforce the concept of *between*
Level:	I or II
Number of players:	6 to 10
Materials:	None

Procedure:	Have three children stand in a line. Then ask the rest of the players to hide their eyes, and while they do so, have the three change position. A player is chosen to guess who is now *between* the other two. If the player is correct, she takes the place on line.

Farm Fun

Purpose:	To reinforce the concepts of *inside, outside, on*
Level:	II or III
Number of players:	4 to 6
Materials:	A bulletin board or flannel board showing a farm yard with a barn and a fence forming a corral Pictures of farm animals
Procedure:	Have children take turns putting the cow in the barn, the chicken on the fence, the horse inside the corral, and so on.
Variation:	Make a gate on the fence, a door on the barn. When the gate is closed, the dog is inside; when the gate is opened, since the dog is now free, there is no longer an "inside."

Blow a Balloon

Purpose:	To teach the concept of relative position
Level:	III
Number of players:	1 to 3
Materials:	A balloon upon which has been drawn a simple closed curve. Inside the curve is a red dot, outside are several green dots.

Procedure: Question the children about the positions of the dots if the balloon is deflated or distorted. Help them to see that no matter what happens, the red dot remains inside the curve while the green dots remain outside.

Silly Names

Purpose: Practice identifying open curves
Levels: III
Number of players: 1 to 3
Materials: Card with three rows: the first showing several open curves, the second, several closed curves, and the third, some open, some closed curves.

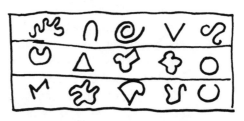

Procedure: Allow one player to give the top row of figures a silly name, and say, "These are _____ but these from the second row are not _____." Children decide which of the bottom figures are _____.

Variation:

Find a Partner

Purpose: To practice distinguishing between open curves, simple closed curves, and curves with two insides

Level: III

Number of players: 6

Materials: Six posters, on each of which you have drawn a curve.

Procedure: Allow each child to choose a curve and find a partner. Have children color the inside of their curves and help them conclude that the first two have one inside, the second two have no inside and the last two have two insides.

In or Out

Purpose: To reinforce the concepts of *inside* and *outside*

Level: III

Number of players: 2 to 4

Materials: A small dot cutout and a large maze drawn on the chalkboard using a simple closed curve.

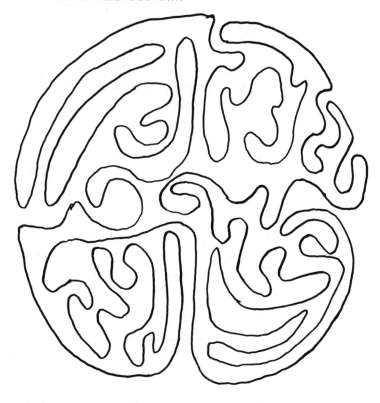

Procedure: Place the cutout somewhere on the maze and ask a player to see if she can follow a path to determine if the doll is inside or outside.

Alphabet Math

Purpose: To reinforce the concept of simple closed curves

Level: III

Number of
players: Small group

Materials: Paper and pencils

Procedure: When teaching manuscript capital letters have the children find those that are simple closed curves (D, O), those that are open curves (C, I, J, L, M, N, S, U, V, W, Z).

Find a Shape

Purpose: To foster awareness of shapes
Level: II or III
Number of players: Small group
Materials: Various geometric solids (many of these are in a set of kindergarten blocks)

cube

rectangular prism

triangular prism

cylinder

sphere

cone

triangular pyramid

rectangular pyramid

Procedure: Choose a block and discuss its shape. Ask: "Does it have any "points" (vertices)? How many? How many flat surfaces (faces)? What are the shapes of the faces? Can you count the edges?" Allow children to handle the block. Next, ask the children to find a

Examining new shapes.

Finding similar shapes.

Finding similar shapes.

159

"shape like this" in the room. As objects are identified, similarities and differences are discussed.

Variation: Using the same blocks, ask children to recall something they have at home or saw on the way to school that was shaped the same.

Finding Partners

Purpose: Foster awareness of shapes; similarities, differences

Level: III

Number of players: Small group

Materials: Geometric solids

Procedure: Display blocks. Ask a child to choose two blocks which he thinks would make good partners. Possible responses include: the sphere and the cylinder because they both roll, the cone and the cylinder because each contain a circular face, the triangular prism and the rectangular prism because one is half as large, the cone and a pyramid because both are the same height (or both come to a point).

I Spy

Purpose: Foster awareness of shapes

Nevel: III

Number of players: Small group

Materials: Geometric solids

Procedure: Choose one player to begin. Have her choose a block and say, "I spy something shaped like this" while whispering her choice to you. The other players take turns asking questions that can be answered "yes" or "no." The child who identifies the object is the next leader.

Find A Shape

Purpose: Practice naming plane shapes: *square, triangle*

Level: I

Number of players: 2 or 3

Materials: Square and triangle shapes from the attribute blocks

Procedure: Allow children to sort the blocks into two piles according to shape. Discuss differences. Identify by name. For reinforcement, put a few blocks into a paper bag or feel box. Allow each player a turn to feel a block and attempt to identify it by name.

Variation: As children become familiar with the shapes, more may be included.

Match 'Em

Purpose: Practice identifying plane figures

Level: I

Number of players: 3 or 4

Materials: Various shapes (and sizes) cut from oaktag
A bulletin board on which the shapes have been traced

Procedure: Have children take turns choosing a shape and matching it with its outline.

Tangrams

Purpose: Practice matching plane shapes

Level: II or III

Number of
players: Small group
Materials: Tangrams (these are available commercially or may be made by enlarging the following diagram)

Various puzzle sheets on which some (or all) of the pieces have been traced

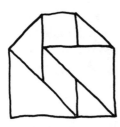 

Procedure: Have children cover the puzzle with the matching tangram pieces.
Variation: Use puzzles with only the outlines shown.

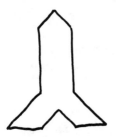 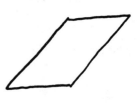

Finger Painting

Purpose: To reinforce the concept of plane shapes

Level: I, II or III

Number of
players: Small group

Materials: Finger paint
Patterns made by cutting bottom from a milk container (square), small cereal box (rectangle,) juice can (circle), and the corner of a shoe box (triangle).

Procedure: Allow children to make designs by dipping shapes into paint and pressing on paper.

Mobiles

Purpose: To reinforce children's knowledge of plane shapes

Level: II or III

Number of
players: Small group

Materials: Yarn or shoelaces, separators cut from straws, various triangle, square, rectangle and circle shapes cut from construction paper

Procedure: Have children string shapes for colorful mobiles.

Baking Cookies

Purpose: To reinforce knowledge of plane shapes

Level: II or III

Number of
players: Small group

Materials:	Premixed cookie dough
	Cookie cutters in the shapes being reinforced
	Oven
Procedure:	Allow children to press the dough into the cookie cutters. After baking, the children enjoy "eating the circle shape."

Pictures from Shapes

Julie, age 4, makes a wagon with shapes.

Purpose:	To reinforce knowledge of plane shapes
Level:	II or III
Number of players:	Small group
Materials:	Circle, square, rectangle, triangle shapes
Procedure:	Allow children to compose a picture using the shapes.

Chapter 14

IN CONCLUSION

The best mathematics program for any child is the one constructed to his or her specific needs and abilities. Activities which only rehash previously conceptualized ideas are too easy. They quickly become boring and provide no growth. On the other hand, children who are expected to master topics for which their internal logical development is not prepared face the same negative results. For this reason, this book stops here. Many areas which may seem to be natural outgrowths of the included activities require a development of logical thought, "a readiness," which preschoolers have not reached. For example, if addition and subtraction are introduced at the optimum time, at the precise time that children perceive all of the interrelationships, they will "learn" the facts in a fraction of the time it would take to memorize each. And it will be a far more pleasurable experience. Maturity of thought is needed to understand the structure of our decimal number system. A child can be taught to parrot by rote that the 6 in 64 stands for tens and that after 60 comes 70. Superficial learning is at best, worthless and, at most, harmful. It is one of the reasons that so many teenagers and adults avoid mathematics. Teachers must know where to start and when to stop.

REFERENCES

CHAPTER 1

Berson, M., and Sherman, C. Divergent Views on Prekindergarten Teacher Education. *Journal of Educational Leadership,* November, 1976, *34*(2) 143–149.

Education Notes Section. *Phi Delta Kappan,* Jan., 1978, *59*(5) 369.

Elkind, D. Early Childhood Education: A Piagetian Perspective. *National Elementary Principal.* Sept., 1971, *51*:48–55.

Hess, R., and Croft, D. *Teachers of Young Children.* New York: Houghton Mifflin, 1972.

Hymes, J. *Teaching The Child Under Six.* Columbus, Ohio: Charles E. Merrill, 1968.

Levenson, D. Whatever Happened To Early Childhood Education? *Instructor,* October, 1977, *84*(3) 67, 72, 134, 135, 138.

Nuffield Mathematics Project. *Mathematics Begins.* New York: John Wiley and Sons, 1970.

CHAPTER 2

Copeland, R. *How Children Learn Mathematics: Teaching Implications of Piaget's Research.* New York: Macmillan, 1979.

Flavell, J. *The Developmental Psychology of Jean Piaget.* Princeton, N.J.: D. Van Nostrand, 1963.

Gruber, H., and Voneche, J. (Eds.). *Essential Piaget.* New York: Basic Books, 1977. (This is an excellent resource text containing the central excerpts from many of Piaget's writings. The editors provide informative summaries before each excerpt.)

Piaget, J. *Six Psychological Studies,* (D. Elkind, Ed.). New York: Random House, 1968.

Piaget, J., and Inhelder, B. *The Psychology of the Child,* New York: Basic Books, 1969.

Wadsworth, B. *Piaget's Theory of Cognitive Development.* New York: Longman, 1971.

CHAPTER 3

Gruber, H., and Voneche, J. (Eds.). *Essential Piaget.* New York: Basic Books, 1977.

Gruen, G. E. Experiences Affecting the Development of Number Conservations in Children. *Child Development,* 1965 *36*:964–979.

Piaget, J. *Six Psychological Studies,* (D. Elkind, Ed.). New York: Random House, 1968.

Wadsworth, B. *Piaget's Theory of Cognitive Development.* New York: Longman, 1971.

CHAPTER 4

Beck, K. Piaget Does Not Live On 'Sesame Street.' *Educational Technology,* 1977 , *53*:610–619.

Elkind, D. Early Childhood Education: A Piagetian Perspective. *National Elementary Principle.* Sept., 1971, *51*:48–55.

Furth, H., and Wachs, H. *Piaget's Theory in Practice: Thinking Goes to School.* New York: Oxford University Press, 1974.

Gruber, H, and Voneche, J. (Eds). *Essential Piaget.* New York: Basis Books, 1977.

Kamii, C. Pedagogical Principles Derived from Piaget's Theory: Relevance for Educational Preactice. In *Piaget in the Classroom,* (M. Schwebel and J. Raph, Eds.) New York: Basic Books, 1973, 199–215.

Piaget, Jean. *Science of Education and the Psychology of the Child.* New York: Orion, 1970.

CHAPTER 5

Gagne, R. M. "Defining Objectives For Six Varieties of Learning," (a taped presentation). Washington: American Educational Research Association, 1971.

Galloway, C. *Psychology For Learning and Teaching.* New York: McGraw Hill, 1976.

Higgins, J. *Mathematics: Teaching and Learning.* Worthington, Oh. Charles A. Jones, 1973.

Skinner, B. F. "Reflections on a Decade of Teaching Machines." *Teachers College Record,* November, 1963, 168–177.

Skinner, B. F. *The Technology of Teaching.* New York: Appleton-Century-Crofts, 1968.

Wadsworth, B. *Piaget's Theory of Cognitive Development.* New York: Longman, 1971.

CHAPTER 6

Copeland, R. *How Children Learn Mathematics: Teaching Implications of Piaget's Research.* New York: Macmillan, 1979.

Piaget, J. *Genetic Epistemology.* New York: Columbia University Press, 1970.

Piaget, J. "How Children Form Mathematical Concepts." *Scientific American,* November, 1953.

SUGGESTED READINGS

CHAPTER 1

Burton, G. "Helping Parents Help Their Preschool Children." *Arithmetic Teacher,* May, 1978, 12–14.

Woods, R. "Preschool Arithmetic Is Important." *Arithmetic Teacher,* January, 1968, *15,* 7–9

CHAPTER 2

Belkin, G., and Gray, J. *Educational Psychology: An Introduction.* Dubuque, Iowa: Wm. C. Brown, 1977.

Lovell, K. *The Growth of Understanding in Mathematics: Kindergarten through Grade Three.* Toronto, Ontario: Holt, Rinehart, and Winston, 1971.

CHAPTER 3

Ginsburg, H. and Opper, S. *Piaget's Theory of Intellectual Development: An Introduction.* Englewood Cliffs, New Jersey: Prentice-Hall, Inc., 1969.

Sigel, E. and Hooper, F. *Logical Thinking in Children: Research Based on Piaget's Theory.* New York: Holt, Rinehart and Winston, 1968.

CHAPTER 4

Bruner, J. *The Process of Education.* Cambridge, MA.: Harvard University Press, 1961.
Wadsworth, B. *Piaget For The Classroom Teacher.* New York: Longman, 1978.

CHAPTER 5

Gagne, R. M. *The Conditions of Learning.* New York: Holt, Rinehart and Winston, 1977.
Strauss, S. "Learning Theories of Gagne and Piaget: Implications For Curriculum Development." *Teachers College Record.* September, 1972, *74* : 81–102.

CHAPTER 6

Furth, H. G. *Piaget For Teachers.* Englewood Cliffs, New Jersey: Prentice-Hall, 1969.
Liedtke, W., and Nelson, L. "Activities in Mathematics For Preschool Children." *Arithmetic Teacher,* November, 1973, 536–541.
Mathematics Learning in Early Childhood. 37th Yearbook. Reston, Virginia: National Council of Teachers of Mathematics, 1975. (Although this resource book goes to grade 3, it does contain imaginative sections on how and what to teach preschoolers.)
Rea, R., and Reys, R. "Mathematical Competencies of Entering Kindergarteners." *Arithmetic Teacher,* January, 1970, *17*:65–74.

READ ALOUD BOOKS

Adler, I. & R. *Sets and Numbers for the Very Young.* Chicago: Day, 1969.

Allen, R. *Numbers—A First Counting Book.* New York: Platt and Munk, 1968.

Berenstain, S. & J. *Inside, Outside, Upside Down.* New York: Random House, 1968.

Brown, M. W. *Four Fur Feet.* Reading, MA.: Addison Wesley, 1961.

Gretz, S. *Teddy Bears 1 to 10.* Chicago: Folett, 1969.

Hall, A. *Let's Count.* Reading, MA.: Addison Wesley, 1976.

Hefter, R. *Shapes.* New York: Larousse, 1976.

Hoban, T. *Shapes and Things.* New York: Macmillan, 1970.

Hoberman, J. & N. *All My Shoes Come in Twos.* Boston: Little Brown, 1957.

Maestro, G. *One More and One Less.* New York: Crown, 1974.

Pienkowski, J. *Numbers.* New York: Harvey House, 1975.

Pienkowski, J. *Sizes.* New York: Harvey House, 1975.

Russel, S. P. *Like and Unlike: A First Look at Classification.* New York: Henry Z. Walck, 1973.

Scott, A. H. *Not Just One.* New York: Lothrop, Lee & Shepard, 1968.

Slobodkin, L. *One is Good but Two are Better.* New York: Vanguard, 1956.

Sugita, Y. *Good Night 1, 2, 3.* New York: Scroll, 1971.

Warner, J. *The Fuzzy Duckling.* Racine, WI.: Golden Press, 1949.

Appendix I

MATERIALS

COMMERCIAL MATERIALS ESSENTIAL IN PRESCHOOL
(OBTAINABLE FROM ANY SUPPLIER)

Attribute blocks
Cuisenaire rods
Geometric solids
Color cone
Beads for stringing
Meter stick
Liter jar
Blank playing cards
Tangrams

EASILY MADE MATERIALS

Feel box
Sorting box
Flannel board and illustrations
Mounted magazine pictures

RECORD KEEPING

A simple (✓) or date when skill is mastered will assist the teacher in planning lessons.

Name	Classification			Ordering			Counting			Etc.
	I	*II*	*III*	*I*	*II*	*III*	*I*	*II*	*III*	...

INDEX